Biochemistry

NOTICE

Medicine is an ever-changing science. As new research and clinical experience broaden our knowledge, changes in treatment and drug therapy are required. The editors and the publisher of this work have made every effort to ensure that the drug dosage schedules herein are accurate and in accord with the standards accepted at the time of publication. Readers are advised, however, to check the product information sheet included in the package of each drug they plan to administer to be certain that changes have not been made in the recommended dose or in the contraindications for administration. This recommendation is of particular importance in regard to new or infrequently used drugs.

Biochemistry:

PreTest® Self-Assessment and Review
Second Edition

Edited by
Ian D. K. Halkerston, Ph. D.

Associate Professor of Biochemistry
University of Massachusetts Medical School
Worcester, Massachusetts

McGraw-Hill Book Company
Health Professions Division
PreTest Series

New York St. Louis San Francisco
Auckland Bogotá Guatemala Hamburg
Johannesburg Lisbon London Madrid
Mexico Montreal New Delhi Panama
Paris São Paulo Singapore Sydney
Tokyo Toronto

Library of Congress Cataloging in Publication Data
Main entry under title:

Biochemistry: PreTest self-assessment and review.

First ed. (1976) edited by J. M. Kirkwood.
Bibliography: p.
1. Biological chemistry--Examinations, questions,
etc. I. Halkerston, Ian D. K.
QP518.5.B58 1979   612'.0076   79-83718
ISBN 0-07-050963-8

3 4 5 6 7 8 9 0 HUHU 8 7 6 5 4 3 2 1 0

Editors: *Mary Ann C. Sheldon, John H. Gilchrist*
Editorial Assistant: *Donna Altieri*
Production Supervisor: *Susan A. Hillinski*
Production Assistants: *Rosemary J. Pascale, Judith M. Raccio*
Designer: *Robert Tutsky*
Printer: *Hull Printing Company*

Contents

Introduction

Biochemistry: PreTest Self-Assessment and Review has been designed to provide medical students, as well as physicians, with a comprehensive and convenient instrument for self-assessment and review within the field of biochemistry. The 500 questions provided have been designed to parallel the format and degree of difficulty of the questions contained in Part I of the National Board of Medical Examiners examinations, the Federation Licensing Examination (FLEX), the Visa Qualifying Examination, and the ECFMG examination.

Each question in the book is accompanied by an answer, a paragraph explanation, and a specific page reference to either a current journal article, a textbook, or both. A two page bibliography, listing all the sources used in the book, follows the last chapter.

Perhaps the most effective way to use this book is to allow yourself one minute to answer each question in a given chapter; as you proceed, indicate your answer beside each question. By following this suggestion, you will be approximating the time limits imposed by the board examinations previously mentioned.

When you finish answering the questions in a chapter, you should then spend as much time as you need verifying your answers and carefully reading the explanations. Although you should pay special attention to the explanations for the questions you answered incorrectly, you should read **every** explanation. The authors of this book have designed the explanations to reinforce and supplement the information tested by the questions. If, after reading the explanations for a given chapter, you feel you need still more information about the material covered, you should consult and study the references indicated.

This book meets the criteria established by the AMA's Department of Continuing Medical Education for up to 22 hours of credit in category 5D for the Physician's Recognition Award. It should provide an experience that is instructive as well as evaluative; we also hope that you enjoy it. We would be very happy to receive your comments.

Amino Acids, Proteins, and Enzymes

DIRECTIONS: Each question below contains five suggested answers. Choose the **one best** response to each question.

1. The structural component of proteins that makes the greatest contribution to their optical absorbance at 280 nm is the

(A) indole ring of tryptophan
(B) phenol ring of tyrosine
(C) benzyl ring of phenylalanine
(D) sulfur atom of cysteine
(E) peptide bond

2. Rotation of polarized light is caused by solutions of all the following amino acids EXCEPT

(A) alanine
(B) glycine
(C) leucine
(D) serine
(E) valine

3. The amino acid that has an ionizable side-chain pK closest to physiologic pH is which of the following?

(A) Cysteine
(B) Glutamic acid
(C) Glutamine
(D) Histidine
(E) Lysine

4. All the following statements about the peptide glu-his-arg-val-lys-asp are true EXCEPT that

(A) it migrates toward the anode at pH 12
(B) it migrates toward the cathode at pH 3
(C) it migrates toward the cathode at pH 5
(D) it migrates toward the cathode at pH 11
(E) its isoelectric point is approximately pH 8

5. Which of the following amino acids has a net positive charge at physiologic pH?

(A) Cysteine
(B) Glutamic acid
(C) Lysine
(D) Tryptophan
(E) Valine

1

6. The graph below shows a titration curve of a common biochemical compound. All the following statements about the graph are true EXCEPT that

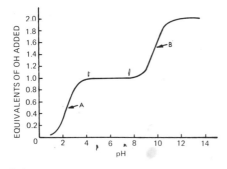

(A) the compound has two ionizable functions

(B) the compound is a simple amino acid

(C) the maximum buffering capacity of the compound is between pH 5 and 7

(D) point A could represent the range of ionization of a carboxyl function

(E) points A and B represent the respective pKs of an acidic and basic function

7. The concentration of hydrogen ions in a solution is expressed as the pH, which is numerically equivalent to

(A) $\log_{10} [H^+]$

(B) $- \log_{10} [H^+]$

(C) $\log_e [H^+]$

(D) $- \log_e [H^+]$

(E) $1/\log_{10} [H^+]$

8. Ninhydrin reacts with amino acids by causing

(A) dehydrogenation

(B) a reduction of the amino group

(C) cleavage of certain peptide bonds

(D) cleavage of the side chain

(E) oxidative decarboxylation

9. Tyrosine may be detected in a mixture of amino acids by which of the following tests?

(A) Sakaguchi reaction (alpha-naphthol and sodium hypochlorite)

(B) Nitroprusside reaction

(C) Ehrlich's reaction (p-dimethylaminobenzaldehyde)

(D) Millon reaction [$Hg(NO_3)_2$ in HNO_3]

(E) Sullivan reaction

10. Which of the following amino acids is synthesized only after incorporation of a precursor into a polypeptide?

(A) Proline

(B) Lysine

(C) Hydroxyproline

(D) Glutamate

(E) Serine

11. Which of the following amino acids is formed by transamination of a member of the citric acid cycle?

(A) Alanine

(B) Lysine

(C) Serine

(D) Aspartic acid

(E) Valine

12. Which of the following amino acids can be transaminated to oxaloacetic acid?

(A) Glutamic acid
(B) Alanine
(C) Threonine
(●) Aspartic acid
(E) Proline

13. Thyroxine is a derivative of

(A) threonine
(B) tryptophan
(●) tyrosine
(D) thiamine
(E) tyramine

14. Which of the following amino acids is NOT essential in mammals?

(A) Phenylalanine
(B) Lysine
(●) Tyrosine
(D) Leucine
(E) Methionine

15. Aspartic acid is incorporated in the synthesis of

(A) porphyrins
(B) steroids
(C) sphingolipids
(●) pyrimidines
(E) coenzyme A

Questions 16-17

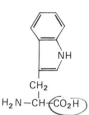

16. The figure shown above is

(A) coenzyme A
(B) NAD
(C) tyrosine
(D) histidine
(●) tryptophan

17. The compound shown above is found in

(A) DNA
(B) mRNA
(●) hemoglobin
(D) sphingomyelin
(E) the citric acid cycle

18. Which of the following amino acids is ketogenic but not glucogenic?

(A) Isoleucine
(B) Tyrosine
(●) Leucine
(D) Phenylalanine
(E) Threonine

19. Which of the following amino acids can contribute carbon to succinyl CoA?

(●) Isoleucine
(B) Leucine
(C) Arginine
(D) Histidine
(E) Tryptophan

20. Which of the following amino acids can undergo direct deamination to 2-keto acids?

(A) Leucine
(B) Threonine
(C) Proline
(D) Isoleucine
(E) Alanine

21. Phenylalanine and tyrosine enter the citric acid cycle after degradation to

(A) pyruvate
(B) fumarate
(C) succinyl-CoA
(D) α-ketoglutarate
(E) citrate

22. N^5-Methyltetrahydrofolate is a methylating agent that transfers a methyl group to

(A) acetate
(B) homocysteine
(C) norepinephrine
(D) pyruvic acid
(E) testosterone

23. S-Adenosylmethionine is a methylating agent that transfers a methyl group to

(A) acetate
(B) homocysteine
(C) norepinephrine
(D) pyruvic acid
(E) testosterone

24. The amino acid that is a major source of carbon for the one-carbon pool is

(A) proline
(B) serine
(C) glutamic acid
(D) threonine
(E) tyrosine

25. Which of the following amino acids can serve as a precursor of the neuro-humoral agent serotonin?

(A) Tyrosine
(B) Proline
(C) Tryptophan
(D) Threonine
(E) Serine

26. Which of the following is a sulfur-containing amino acid NOT found in proteins?

(A) Homocysteine
(B) Cysteine
(C) Methionine
(D) Cystine
(E) Threonine

27. Which of the following amino acid residues in a protein can undergo phosphorylation-dephosphorylation reactions during activation-deactivation cycles in activity of the protein?

(A) Aspartic acid
(B) Proline
(C) Leucine
(D) Serine
(E) Glycine

28. In urea synthesis, which of the following reactions requires adenosine triphosphate (ATP)?

(A) Arginine → ornithine + urea
(B) Oxaloacetate + glutamate → aspartate + α-ketoglutarate
(●) Citrulline + aspartate → argininosuccinate
(D) Fumarate → malate
(E) None of the above

29. In the cycle shown below, compound B is

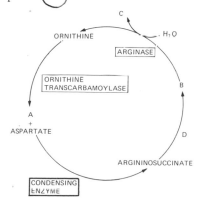

(A) pyruvate
(B) urea
(C) malate
(D) lactate
(●) arginine

30. The decarboxylation of which of the following amino acids produces a vasodilating compound?

(A) Arginine
(B) Aspartic acid
(●) Histidine
(D) Glutamine
(E) Proline

31. All the following statements about the α-helix are true EXCEPT that

(A) it is stabilized by intramolecular hydrogen bonds
(B) it is stabilized by minimizing unfavorable R-group interactions
(●) it is stabilized by hydrophobic interactions
(D) it is one type of secondary structure found in some proteins
(E) prolyl and glycyl residues tend to interrupt α-helical structure

32. Fibroblasts secrete collagen in the form of

(A) a triple-stranded helical rod having a molecular weight of 285,000
(B) a triple-stranded helical rod having a molecular weight of 285,000, associated with a polysaccharide
(●) a triple-stranded helical rod having a molecular weight of 360,000
(D) a triple-stranded polypeptide structure having a molecular weight of 95,000
(E) none of the above

33. Lactate dehydrogenase (LDH) is a tetramer composed of two different polypeptide chains. Assuming these chains associate at random to form the enzyme, how many isozymes does this enzyme possess?

(A) Two
(B) Three
(C) Four
(D) Five
(E) Six

Questions 34-35

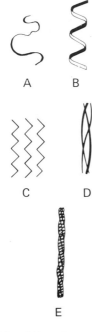

A   B

C   D

E

34. Which of the schematic drawings of protein configurations shown above represents a super coiled helix?

(A) Figure A
(B) Figure B
(C) Figure C
(D) Figure D
(●) Figure E

35. Figure D is a three-stranded structure which could represent the conformation of

(A) hemoglobin
(B) α-keratin
(C) polylysine
(D) silk fibroin
(●) tropocollagen

36. All the following gastrointestinal enzymes are secreted as inactive zymogens (proenzymes) EXCEPT

(●) ribonuclease
(B) trypsin
(C) chymotrypsin
(D) carboxypeptidase
(E) pepsin

37. Which enzyme has the greatest specificity for peptide bonds on the carboxyl side of a cationic amino acid side chain?

(A) Carboxypeptidase
(B) Chymotrypsin
(●) Trypsin
(D) Pepsin
(E) Rennin

38. Which of the following statements about myosin is true?

(A) It is a zinc-requiring enzyme
(B) It is a spherically symmetric molecule
(C) It is a cyclic adenosine 5'-monophosphate (AMP) phosphodiesterase
(●) It is an actin-binding protein
(E) It is low in α-helix content

39. Which of the following statements about immunoglobulins is true?

(A) They contain no carbohydrate
(B) They can be separated into different classes by electrophoresis
(C) They maintain a constant serum level
(D) They are synthesized by the mucosal cells of the appendix
(E) Their heavy and light chains are linked by nucleic acids

40. The first step in the catabolism of hemoglobin occurs when hemoglobin

(A) undergoes fission of the α-methene bridge
(B) is converted to biliverdin in the liver
(C) is converted to bilirubin in the reticuloendothelial cells
(D) is conjugated with glucuronic acid in the liver
(E) is reduced in the liver

41. The normal brown-red color of feces results from the presence of

(A) stercobilin
(B) urobilinogen
(C) bilirubin
(D) mesobilirubin
(E) biliverdin

42. K_m and V_{max} can be determined from the Lineweaver-Burk plot of the Michaelis-Menten equation shown below. When V is the reaction velocity at substrate concentration S, the x-axis experimental data are expressed as

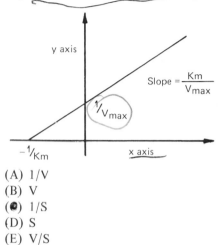

(A) 1/V
(B) V
(C) 1/S
(D) S
(E) V/S

43. If an enzyme is active only when a particular histidyl group is not protonated, then increasing the proton concentration, i.e., decreasing pH, would give which of the following types of inhibition pattern?

(A) Uncompetitive
(B) Noncompetitive
(C) Competitive
(D) Mixed
(E) None of the above

44. A purely competitive inhibitor of an enzyme has which of the following kinetic effects?

(A) Increases K_m without affecting V_{max}
(B) Decreases K_m without affecting V_{max}
(C) Increases V_{max} without affecting K_m
(D) Decreases V_{max} without affecting K_m
(E) Decreases both V_{max} and K_m

45. Enzymes as classic catalysts accomplish which of the following energy effects?

(A) Raise the energy of activation
(B) Lower the energy of activation
(C) Raise the energy level of the products
(D) Lower the energy level of the reactants
(E) Decrease the free energy of the reaction

46. Given that
$$\Delta G° = -2.3RT \log K_{eq},$$
determine the free energy of the following reaction.

A	+	B	⇌	C
10		10		10
moles		moles		moles

(A) −9.2 RT
(B) −4.6 RT
(C) −2.3 RT
(D) +2.3 RT
(E) +4.6 RT

47. The K_m of the enzyme giving the kinetic data shown below is

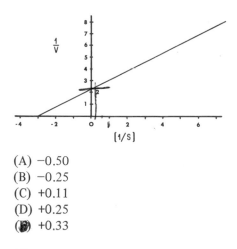

(A) −0.50
(B) −0.25
(C) +0.11
(D) +0.25
(E) +0.33

48. If curve X in the graph shown below represents no inhibition for the reaction of some enzyme with its substrate, which of the other curves would represent competitive inhibition of the same reaction?

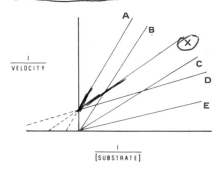

(A) A
(B) B
(C) C
(D) D
(E) E

49. If an enzyme behaves according to classic Michaelis-Menten kinetics, from a double-reciprocal plot of velocity versus substrate concentration, the value for the Michaelis constant (K_m) of the substrate can be determined graphically as the

(A) point of inflection of the curve
(B) slope of the curve
(C) absolute value of the intercept of the curve with the x axis
(D) reciprocal of the absolute value of the intercept of the curve with the x axis
(E) reciprocal of the absolute value of the intercept of the curve with the y axis

50. The overall energy changes in biochemical reactions are

(A) influenced by the energy barrier of the reaction
(B) altered by cofactors
(C) proportionate to the concentration of the reactants
(D) significant at equilibrium
(E) independent of the mechanism of the reaction

51. Which of the following oxidation-reduction systems has the highest redox potential?

(A) Fumarate/succinate
(B) Ubiquinone ox/red
(C) Fe^{+++} cytochrome a/Fe^{++}
(D) Fe^{+++} cytochrome b/Fe^{++}
(E) NAD^+/NADH

52. The second law of thermodynamics states that

(A) perpetual motion is theoretically attainable at $0°K$
(B) energy and mass are conserved and interchangeable
(C) in an energetically closed system any process goes spontaneously from a state of lesser order to a state of greater order
(D) in an energetically closed system any process goes spontaneously in the direction of increased entropy
(E) any system changes spontaneously in the direction of decreased free energy

53. The plant poison atractyloside has the ability specifically to

(A) inhibit the interaction of cytochrome a_3 with molecular oxygen
(B) inhibit the facilitated exchange diffusion of ATP and ADP across the mitochondrial inner membrane
(C) uncouple oxidative phosphorylation
(D) block the interaction of NADH dehydrogenase with coenzyme Q
(E) block the interaction of cytochrome c with the cytochrome $a + a_3$ complex

54. Dinitrophenol would be most likely to inhibit cell function by disrupting

(A) glycolysis
(B) hepatic gluconeogenesis
(C) oxidative phosphorylation
(D) the citric acid cycle
(E) none of the above

55. Polymyxin is unique among chemotherapeutic agents because it is bactericidal in the absence of cell growth. It exerts its effect by

(A) binding to DNA as an insertion mutation
(B) binding to DNA polymerase
(C) binding to polysome-bound mRNA
(D) detergent-like disruption of membranes
(E) dissociating sigma from RNA polymerase

56. Cycloserine inhibits transpeptidation in the formation of the peptidoglycan cell wall network of gram-positive organisms. This action of cycloserine is competitively inhibited by

(A) D-alanine
(B) D-serine
(C) D-glutamine
(D) L-lysine
(E) diaminopimelic acid

57. The activity of most single polypeptide enzymes can be represented by the hyperbolic curve A shown below. However, the activity of homotropic regulatory enzymes shows a sigmoid dependence on substrate concentration, curve B. This sigmoid relationship between substrate concentration and reaction velocity indicates that

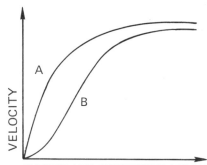

(A) homotropic enzymes are polymers
(B) homotropic enzymes must catalyze several separate reactions on the way to the final product
(C) homotropic enzymes catalyze reactions more slowly than single polypeptide enzymes
(D) the reaction rate is independent of substrate concentration
(E) the binding of one substrate molecule enhances subsequent substrate binding and activity

58. Some antibiotics act as ionophores, which means that they

(A) interfere directly with bacterial
cell-wall synthesis
(B) have a detergent-like effect on cell
membranes
(●) increase cell-membrane permeability to specific ions
(D) inhibit both transcription and
translation
(E) inhibit only translation

59. Which of the following types of
enzymes does NOT have a demon-
strable inducer?

(A) Allosteric enzyme
(●) Constitutive enzyme
(C) Isozymic enzyme
(D) Inhibited enzyme
(E) Cooperative enzyme

60. *De novo* synthesis of an enzyme,
promoted by the substrate on which
it acts, is characterized by the term

(A) activation
(B) derepression
(C) gratuity
(●) induction
(E) constitutivity

61. Control of metabolic pathways
may be exerted by enzyme repression
or induction. In vertebrates, this form
of enzyme control occurs primarily
in the

(A) heart
(●) liver
(C) brain
(D) skeletal muscle
(E) bone

62. A hypothetical biosynthetic pathway is shown in the diagram below. A
microbial organism defective in one
enzyme of this path is grown in a
medium containing X. Large amounts
of M and L are found in the organism
but none of Z. In which enzyme is the
mutation expressed?

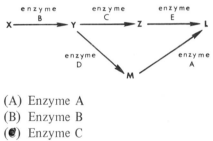

(A) Enzyme A
(B) Enzyme B
(●) Enzyme C
(D) Enzyme D
(E) Enzyme E

DIRECTIONS: Each question below contains four suggested answers of which **one** or **more** is correct. Choose the answer:

A	if	1, 2, and 3	are correct
B	if	1 and 3	are correct
C	if	2 and 4	are correct
D	if	4	is correct
E	if	1, 2, 3, and 4	are correct

63. Which of the following statements about most peptide bonds between L-amino acids in proteins are true?

(1) The peptide bond has partial double-bond character

(2) The peptide bond is shorter than a normal carbon-carbon single bond

(3) The peptide bond usually has a *trans* configuration with respect to the α-carbons of the two amino acids involved in the peptide bond

(4) There is perfectly free rotation about the peptide bond

64. Which of the following amino acids have relatively nonpolar, hydrophobic side chains?

(1) Isoleucine

(2) Methionine

(3) Proline

(4) Phenylalanine

65. Which of the following statements about amino acid biosynthesis are true?

(1) Essential amino acids are required in an animal's diet to maintain a proper nitrogen balance

(2) Methionine is an essential amino acid for humans

(3) The "essentiality" of an amino acid can vary as a function of physiologic state, e.g., pregnancy

(4) Tyrosine is an essential amino acid for humans

66. In mammalian tissues, glycine is a biosynthetic precursor of

(1) heme

(2) creatine

(3) guanine

(4) thymine

67. In mammalian tissues, serine can be a biosynthetic precursor of

(1) methionine

(2) glycine

(3) tryptophan

(4) choline

68. Penicillin may be considered to be a condensation product of

(1) alanine
(2) cysteine
(3) isoleucine
(4) valine

69. D-Alanine, illustrated below, is a structural analog of

(1) streptomycin
(2) vancomycin
(3) bacitracin
(4) cycloserine

70. Which of the following drugs work by decreasing the levels of folate co-factors in bacteria?

(1) *p*-Aminosalicylic acid
(2) Isoniazid
(3) Sulfisoxazole (Gantrisin)
(4) Tetracycline

71. Which of the following statements about the structure of proteins are true?

(1) Intrachain disulfide bonds may not be crucial in predetermining the conformation of a protein molecule
(2) Charged amino acid side chains tend to be on the outside of the molecule exposed to solvent
(3) The primary structure of proteins is among the more important factors in determining the higher order structure
(4) Hydrophobic side chains of amino acid residues only rarely are buried within molecules, thus protected from water

72. Which of the following statements about gamma immunoglobulins are true?

(1) Immunoglobulins have two antigen binding sites per antibody molecule
(2) In multiple myeloma, incomplete immunoglobulins may be excreted in the urine (Bence Jones proteins)
(3) Both heavy and light immuno-globulin chains have constant C-terminal and variable N-terminal sequences
(4) The only chemical forces holding immunoglobulin protein chains together are noncovalent in nature

73. Which of the following statements concerning trypsinogen and chymotrypsinogen are false?

(1) They have considerable homologies in primary sequence
(2) They can be converted into active enzymes by limited digestion with trypsin
(3) They are secreted by exocrine cells in the pancreas
(4) They are exopeptidases

74. It is true of most enzymes that they

(1) increase the rapidity of the reaction they catalyze
(2) are specific for the substrate as well as the reaction catalyzed
(3) are large polypeptides with a molecular weight greater than 5,000 daltons
(4) are most active near neutral pH

75. The figure below shows the structure of an immunoglobulin hydrolyzed by papain to form two A fragments and one B fragment. It is true of fragment A that it

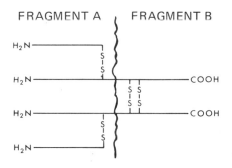

FRAGMENT A   FRAGMENT B

(1) contains the antibody-combining site
(2) is Bence Jones protein
(3) contains the "hypervariable region" sequences
(4) is the heavy chain

76. Proteins that contain a porphyrin ring include

(1) hemoglobin
(2) myoglobin
(3) cytochrome
(4) catalase

77. Which of the following proteins contain iron?

(1) Cytochrome c
(2) Hemoglobin
(3) Myoglobin
(4) Peroxidase

78. Which of the following statements about the complement system are true?

(1) It is composed of a group of soluble globulins
(2) It is a self-assembling system
(3) It is lytic to red blood cells
(4) It usually requires only a bacterium to initiate its action

79. The Monod model for allosteric enzymes assumes that

(1) all allosteric enzymes are polymers
(2) each subunit bears both a catalytic and an allosteric site
(3) the binding affinity of a subunit for a ligand may differ according to the conformational state of the subunit
(4) transitions from state to state involve changes in all subunits

80. If succinic acid (fumarate/succinate redox potential, + 0.03 volts) is added to an equilibrium mixture of ferric and ferrous sulfate (ferric/ferrous redox potential, + 0.77 volts),

(1) the concentration of ferric sulfate will increase
(2) the concentration of ferric sulfate and fumarate will increase
(3) there will be no change in the ratio of ferric to ferrous forms
(4) the concentration of ferrous sulfate and fumarate will increase

Questions 81-82

The velocity-substrate curve below characterizes an allosteric enzyme system.

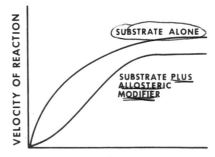

81. Which of the following statements are true?

(1) A modifier present at the allosteric site can also affect the catalytic site
(2) A modifier changes the binding constant for the substrate but not the velocity of the reaction
(3) The binding of substrate is dependent on its concentration
(4) The binding of modifier is independent of its concentration

82. The type of allosteric behavior exhibited above is characteristic of

(1) phosphoglucose isomerase
(2) aspartate transcarbamoylase
(3) lactate dehydrogenase
(4) hemoglobin

83. Which of the following statements about hemoglobin transport of O_2 are true?

(1) Each of the four heme moieties binds O_2 independently
(2) The graph of percent O_2 bound versus O_2 pressure is sigmoid in shape
(3) O_2 binds to hemoglobin more avidly than does CO
(4) The binding of O_2 to hemoglobin causes no valence change in the iron of the heme moiety

84. Which of the following hypotheses are currently being employed to explain oxidative phosphorylation?

(1) Pasteur effect
(2) Chemical coupling hypothesis
(3) Warburg hypothesis
(4) Chemiosmotic hypothesis

85. Cytochrome b of the mitochondrial electron transport chain has which of the following properties?

(1) A standard redox potential lower than that of cytochromes c and a
(2) Ready detachment from mitochondrial membranes
(3) Low reactivity with cyanide or CO, which are inhibitors of some hemoproteins
(4) Ease of reaction with cytochrome a

86. Which of the following statements about oxidation-reduction potentials are true?

(1) The common standard potential of zero is arbitrarily assigned to the hydrogen electrode
(2) pH has no relation to oxidation-reduction potentials
(3) Free energy changes may be calculated from oxidation-reduction potentials
(4) Metallic electrodes are required for determination of oxidation-reduction potentials

87. Phosphorylation sites in the mitochondrial respiratory chain may occur between

(1) coenzyme Q and cytochrome b
(2) cytochrome b and cytochrome c
(3) pyruvate and NAD
(4) NAD and a flavoprotein

88. Which of the following statements about sickle cell anemia are true?

(1) It results from a single amino acid change in hemoglobin
(2) It is widely distributed in areas of high malaria fatality
(3) It is seen in homozygous individuals only
(4) It is caused by a phospholipid change in erythrocytes

89. Synthesis of which of the following proteins is inducible?

(1) β-Galactoside permease of *B. subtilis*
(2) β-Lactamase in *S. aureus*
(3) Glucose-phosphate transferase of *E. coli*
E (4) Immunoglobulins in cat lymphocytes

90. Which of the following statements about isozymes of a given enzyme are true?

(1) They are composed of distinct multimeric complexes
B (2) They have differing substrate specificities
(3) They may exhibit different K_m values for substrates or cofactors
(4) They usually exhibit identical electrophoretic mobility

91. Factors circulating in the bloodstream that are involved in the formation of blood clots include

(1) plasmin
(2) fibrinogen
C (3) heparin
(4) platelets

92. The active transport of β-galactosides in *E. coli* is characterized by

(1) a requirement for an energy source
(2) a saturating concentration of galactoside above which a more rapid rate of uptake cannot be obtained
(3) a rate of efflux dependent on the intracellular concentration of galactoside
E (4) being inducible in cells grown solely in lactose as an energy source

93. Heavy chains of IgG antibody may be separated from light chains with

(1) ethanolamine
(2) pepsin
(3) papain
D (4) mercaptoethanol

DIRECTIONS: The groups of questions below consist of lettered choices followed by several numbered items. For each numbered item select the **one** lettered choice with which it is **most** closely associated. Each lettered choice may be used once, more than once, or not at all.

Questions 94-95

For each amino acid, choose the compound from which it is synthesized.

(A) Glutamate
(B) Aspartate
(C) Ribose 5-phosphate
(D) Shikimic acid
(E) Serine

C 94. Histidine

A 95. Proline

Questions 96-99

For each amino acid that follows, select the side chain description that is most appropriate.

(A) Acidic
(B) Basic
(C) Aromatic
(D) Sulfur-containing
(E) Branched chain aliphatic

D 96. Cysteine

B 97. Lysine

E 98. Leucine

C 99. Phenylalanine

Questions 100-102

For each amino acid that follows, select the substance in which it appears in highest concentration.

(A) Chondroitin sulfate
(B) Collagen
(C) Keratin
(D) Melanin
(E) Myosin

C 100. Cystine

D 101. Tyrosine

B 102. Hydroxyproline

Questions 103-104

For each chemical cleavage of proteins that follows, select the reagent that produces it.

(A) Cyanogen bromide
(B) Trypsin
(C) 2,4-Dinitrophenol (DNP)
(D) Mercaptoethanol
(E) Chymotrypsin

D 103. Cleavage of protein disulfide bridges

A 104. Splitting of peptide chains at the carboxyl side of methionine residues

Questions 105-107

For each pancreatic enzyme that follows, select the principal peptidolytic action that is most appropriate.

(A) Cleaves after proline
(B) Cleaves after methionine
(C) Cleaves after aromatic amino acids
(D) Cleaves after lysine and arginine
(E) Exopeptidolysis

105. Trypsin

106. Chymotrypsin

107. Carboxypeptidase

Questions 108-110

In the following enzyme-catalyzed reaction,

$$S + E \underset{k_2}{\overset{k_1}{\rightleftharpoons}} S \cdot E \overset{k_3}{\Rightarrow} P + E$$

E, S, and P are the enzyme, substrate, and product, respectively; k_1, k_2, and k_3 are rate constants. For each kinetic parameter that follows, select the expression that best defines it.

(A) $k_3 E_T$ (E_T is the total enzyme concentration)
(B) k_2/k_3
(C) k_2/k_1
(D) $(k_2 + k_3)/k_1$
(E) None of the above

108. V_{max}

109. K_s (dissociation constant for E·S complex)

110. K_m (Michaelis constant)

Amino Acids, Proteins, and Enzymes

Answers

1. The answer is A. *(Lehninger, ed 2. p 83. White, ed 6. p 81.)* The molal extinction coefficients of tryptophan, tyrosine, and phenylalanine at 280 nm are 5,500, 540, and 120 respectively. As most proteins contain considerably more tyrosine residues than tryptophan residues, protein absorption at 280 nm is due chiefly to tyrosine. The absorbance maximum of peptide bonds is at 215 nm, and the disulfide group of cystine absorbs weakly at 240 nm. The sulfur atom of cysteine does not contribute significantly to absorption in this portion of the spectrum.

2. The answer is B. *(White, ed 6. p 82.)* Because glycine has two hydrogen atoms bound to its α-carbon, it is without right-handed and left-handed forms and, therefore, is optically inactive. All other amino acids found in proteins are optically active and have the L-configuration, which refers to absolute configuration and has been shown to be related to L-glyceraldehyde. Moreover, amino acids, which may be dextro- (d-) or levo- (l-) rotatory, display a strong shift of rotation of plane-polarized light with increasing hydrogen ion concentration.

3. The answer is D. *(White, ed 6. pp 106-107.)* Proteins that contain many ionizable groups are effective intracellular buffers. The imidazolium group of histidine has a pK of approximately 6.5. Histidine is the only amino acid with a side chain pK approximately at neutrality.

4. The answer is D. *(White, ed 6. pp 106-107.)* At pH 11 or 12 the peptide glu-his-arg-val-lys-asp has a net negative charge and, as an anion, moves to the anode. At pH 3 or 5, the peptide has a net positive charge and, as a cation, moves to the cathode. The neutral ion species (isoelectric form) is formed when the proton of the imidazolium group of histidine has completely dissociated (approximately pH 8).

5.   The answer is C. *(White, ed 6. pp 77-89.)* At physiologic pH, the epsilon-amino group of lysine carries a positive charge. The carboxylate side chain of glutamic acid is negative, and the side chains of the remaining amino acids mentioned in the question are uncharged.

6.   The answer is C. *(Mahler, ed 2. p 14.)* The figure in the question shows the titration curve of glycine, an amino acid with two dissociable protons, one from the α-carboxyl group and the other from the α-amino group. The maximum buffering capacity of any ionizable function is at the pH equivalent to the pK_a of the dissociation, as represented by points A and B on the graph.

7.   The answer is B. *(Harper, ed 16. p 13.)* The acidity of aqueous solutions is expressed as the pH and is equal to the negative logarithm to the base 10 of the hydrogen-ion concentration. Thus for pure water, which has a $[H^+]$ concentration of 1.0×10^{-7} molar, pH $= -\log 10^{-7} = -(-7) = 7.0$.

8.   The answer is E. *(Mahler, ed 2. p 54.)* Ninhydrin reacts with α-amino acids to cause decarboxylation and deamination while it is reduced to hydrindantin. One molecule of each of the products of this oxidative deamination—ammonia, ninhydrin, and hydrindantin—then complex to form a purple pigment with an optical absorbance maximum of about 570 nm.

9.   The answer is D. *(Mahler, ed 2. p 54.)* The Millon reaction [Hg(NO₃)₂ in nitric acid with a trace of nitrous acid] is a test for phenolic compounds. Tyrosine reacts in this test to yield a red color. The Pauly reaction with diazotized sulfanilic acid in alkaline solution would detect tyrosine and histidine as well.

10.   The answer is C. *(Mahler, ed 2. pp 775-777. Stryer, pp 208-209.)* Hydroxyproline is not directly incorporated into polypeptides. Prolyl residues already incorporated in peptide linkage are hydroxylated by a reaction using molecular oxygen and α-ketoglutarate to give hydroxyprolyl residues, which are found exclusively in collagen and related proteins.

11.   The answer is D. *(Lehninger, ed 2. pp 463-465. Stryer, pp 307-308.)* Alanine is the transamination product of pyruvate, which is not an intermediate of the citric acid cycle. Aspartic acid is the transamination product of oxaloacetic acid, which is a member of the citric acid cycle. Serine is formed from glycolytic pathway intermediates. Lysine and valine are dietary essentials and are not synthesized by humans.

12. The answer is D. *(McGilvery, pp 374-376, 384.)* Aspartic acid transaminates with α-ketoglutarate to form oxaloacetate and glutamate. Alanine transaminates to pyruvate, proline gives rise to glutamate, and threonine is directly deaminated to form 2-ketobutyrate, which is metabolized to succinyl coenzyme A (CoA) via propionyl CoA.

13. The answer is C. *(White, ed 6. pp 1196-1199.)* Thyroxine, a derivative of tyrosine, is formed by iodination and coupling of peptide-linked tyrosyl residues of thyroglobulin in the thyroid gland. The hormone contains over a third of the total iodine in the human thyroid.

14. The answer is C. *(White, ed 6. p 689.)* Because the enzyme phenylalanine hydroxylase exists in the normal mammalian liver and kidney, phenylalanine "spares" the dietary requirement for tyrosine. The other amino acids listed in the question must be supplied by diet.

15. The answer is D. *(White, ed 6. p 765.)* Aspartate and carbamoyl phosphate are condensed to give carbamoyl aspartate in one of the early steps of pyrimidine biosynthesis. This reaction is catalyzed by the enzyme aspartate transcarbamoylase.

16. The answer is E. *(Mahler, ed 2. pp 47-49.)* The structure shown in the question is that of the amino acid tryptophan, which has the most bulky side chain of any amino acid involved in protein synthesis. The side chain is hydrophobic, and the aromatic ring allows tryptophan to interact with other aromatic ring amino acids, like phenylalanine, by pi-electron overlaps.

17. The answer is C. *(White, ed 6. pp 750-751.)* The amino acid tryptophan is used sparingly by living organisms as an amino acid residue in proteins. For humans it is an essential amino acid; higher organisms cannot achieve its synthesis.

18. The answer is C. *(Lehninger, ed 2. p 629.)* The carbon atoms of leucine, a purely ketogenic amino acid, cannot be utilized to form glucose. Some carbon atoms from the other amino acids listed in the question, however, can contribute to glucose formation. Thus, threonine is glucogenic, and isoleucine, tyrosine, and phenylalanine are both ketogenic and glucogenic.

19. The answer is A. *(McGilvery, pp 376, 389, 395, 420.)* Isoleucine is degraded to acetyl CoA and propionyl CoA. The latter is converted to methylmalonyl CoA which is isomerized to succinyl CoA. Leucine is degraded to acetoacetate and acetyl CoA; arginine and histidine to α-ketoglutarate via glutamate; and tryptophan to alanine, crotonyl CoA, and formate.

20. The answer is B. *(McGilvery, pp 356-357.)* Threonine undergoes direct deamination to 2-ketobutyrate. This type of reaction is thermodynamically unfavored for most amino acids, which are simple amines, but proceeds readily as an exergonic reaction when a hydroxyl group is present, as in the case of threonine or serine. A similar reaction occurs in glycolysis when glycerate, with an adjacent hydroxyl group, is dehydrated to pyruvate with the liberation of enough energy to support the formation of a high-energy phosphate bond. The remaining amino acids listed in the question lose their nitrogen by transamination with a 2-keto acid, proline being first converted to glutamate.

21. The answer is B. *(White, ed 6. pp 740-741.)* Beginning with its conversion to tyrosine, phenylalanine follows a metabolic pathway that ultimately issues in two products of degradation, fumarate and acetoacetate, leading directly into the citric acid cycle. This pathway involves the intermediate products homogentisic and 4-maleylacetoacetic acids.

22. The answer is B. *(Mahler, ed 2. pp 781, 807.)* Homocysteine is methylated by N^5-methyltetrahydrofolate to generate methionine, which, after conversion to S-adenosylmethionine, can donate the methyl group to other compounds. Transmethylation reactions include the conversion of guanidinoacetate to creatine and of acetylserotonin to melatonin, among others.

23. The answer is C. *(Mahler, ed 2. p 807.)* Norepinephrine transmethylates with S-adenosylmethionine to form epinephrine and S-adenosylhomocysteine. The reaction is catalyzed by norepinephrine methyltransferase.

24. The answer is B. *(McGilvery, pp 416-418.)* The transfer of a single carbon atom of serine to tetrahydrofolate to form 5,10 methylene tetrahydrofolate is an important contribution of carbon to the one-carbon pool. As serine can be derived from glycolytic pathway intermediates, glucose carbon can therefore enter the one-carbon pool. Other important sources of one-carbon fragments are the amino acids histidine and glycine.

25. The answer is C. *(White, ed 6. pp 722-723, 1117.)* Serotonin is formed from tryptophan by hydroxylation to 5-hydroxytryptophan, which is decarboxylated to yield 5-hydroxytryptamine (serotonin). An important neurotransmitter, serotonin is believed to mediate sleep and sensory functions.

26. The answer is A. *(Mahler, ed 2. pp 44-47. White, ed 6. pp 687-688, 694-695.)* Homocysteine is a sulfur-containing amino acid that is not found in proteins. It is an intermediate in the formation of cysteine from the essential sulfur-containing amino acid, methionine.

27. The answer is D. *(White, ed 6. p 229.)* Protein kinases can transfer the gamma-phosphate group of ATP to the hydroxyl group of serine residues in a protein with an accompanying change in the function of the protein. For example, both phosphorylated and dephosphorylated forms of the enzyme glycogen synthase are essential to the control of glycogenesis.

28. The answer is C. *(White, ed 6. pp 698-700.)* Argininosuccinate synthetase, the enzyme that condenses aspartate with citrulline to yield argininosuccinate, also hydrolyzes ATP to adenosine monophosphate (AMP) and pyrophosphate. In the presence of argininosuccinase, argininosuccinic acid readily yields arginine, which is hydrolyzed in the liver to form urea under the catalytic action of arginase. The fumarate formed in the argininosuccinase reaction can be hydrated to malate by fumarase, and the malate oxidized to oxaloacetate by malate dehydrogenase; aspartate, required in the argininosuccinate synthetase reaction, can be formed from oxaloacetate by transamination with glutamate.

29. The answer is E. *(White, ed 6. pp 698-699.)* In the urea cycle, ornithine is carbamoylated to yield citrulline which condenses with aspartate to yield argininosuccinate. Argininosuccinase catalyzes the formation of arginine which in turn is hydrolyzed to form ornithine and urea.

30. The answer is C. *(White, ed 6. p 720.)* Decarboxylation of histidine yields histamine, a compound that is active both as a vasodilator and in the mediation of allergy. Histamine also is used in the clinical evaluation of gastric secretions by virtue of its ability to stimulate the release of pepsin and hydrochloric acid.

31. The answer is C. *(Mahler, ed 2. pp 133-145.)* The α-helix is one type of periodic or secondary structure found in some proteins. Among the stabilizing forces are, a) the intramolecular hydrogen bonds formed between peptide bonds, and b) the geometric arrangement of the helix that minimizes the unfavorable steric interactions among the R groups of the amino acid residues. Hydrophobic interactions are not thought to be important in stabilizing α-helical structure, and regions of the α-helix do tend to be interrupted by prolyl or glycyl residues.

32. The answer is C. *(Stryer, pp 211-217.)* Collagen is secreted by fibroblasts in the form of procollagen (MW 360,000). Each peptide component of the triple-stranded helical rod bears additional nonhelical peptides covalently bonded to N-terminal (and possibly the C-terminal) ends of the chains. These nonhelical regions contain many intrachain disulfide linkages. After secretion, extracellular peptides cleave off the nonhelical regions to yield tropocollagen (MW 285,000).

33. The answer is D. *(White, ed 6. p 448.)* Assuming lactate dehydrogenase (LDH) is a tetrameric protein formed by random association of two electrophoretically distinct polypeptides, A and B, one can predict the existence of five isozymes: A_4, A_3B_1, A_2B_2, A_1B_3, B_4.

34. The answer is E. *(Mahler, ed 2. pp 139-140.)* In the illustration accompanying the question, Figure A is a random coil; Figure B an α-helix; Figure C a pleated sheet; and Figure D a triple-stranded helix. Figure E is a super-coiled helix, one of the forms of hair keratin.

35. The answer is E. *(Mahler, ed 2. pp 139-140.)* Tropocollagen has the structure of a triple helix because many proline and hydroxyproline residues prevent the hydrogen bond formation necessary for an α-helix (Figure B in the illustration accompanying the questions).

36. The answer is A. *(Mahler, ed 2. pp 762-769.)* Pepsinogen, procarboxypeptidase, chymotrypsinogen, and trypsinogen are inactive zymogen precursors of the gastrointestinal enzymes pepsin, carboxypeptidase, chymotrypsin, and trypsin respectively. Ribonuclease is secreted in its active form.

37. The answer is C. *(White, ed 6. pp 139-141, 232, 678-680.)* Trypsin is an endopeptidase that hydrolyzes the carboxyl peptide linkages of arginine and lysine, which are both positively charged amino acids. Chymotrypsin is specific for the carboxyl peptide linkages of aromatic amino acids. Carboxypeptidase hydrolyzes amino acids sequentially from the carboxyl end of proteins. Rennin produces paracasein from casein, allowing its degradation by pepsin into proteoses and peptones.

38. The answer is D. *(White, ed 6. pp 1085-1089.)* Myosin is a rod-shaped protein, regarded as a two-stranded α-helical coil, that binds two molecules of actin to its two globular heads. The heads contain the adenosine triphosphatase (ATPase) activity. Actin and myosin both are essential to vertebrate striated muscle contraction. Myosin does not require zinc for its biologic function.

39. The answer is B. *(White, ed 6. pp 930-932.)* Immunoglobulins contain light and heavy chains linked by disulfide bonds and a small amount of carbohydrate. Immunoglobulins are synthesized by lymphocytes and plasma cells in response to specific antigens and can be differentiated into at least five classes, i.e., IgG, IgA, IgM, IgD, and IgE.

40. The answer is B. *(White, ed 6. pp 986-988.)* The catabolism of hemoglobin first involves scission of the α-methene bridge to yield choleglobin. Removal of the apoprotein yields verdohemochrome. Verdohemochrome is converted to biliverdin which, in turn, is converted to bilirubin.

41. The answer is A. *(White, ed 6. pp 987-988.)* Stercobilin is the pigment which imparts the characteristic color to stools. Bilirubin diglucuronide is hydrolyzed to yield bilirubin which is reduced to urobilinogen by bacterial flora. Urobilinogen is further reduced to stercobilinogen. Oxidation of stercobilinogen in air yields stercobilin.

42. The answer is C. *(Mahler, ed 2. p 277.)* In the Lineweaver-Burk transformation of the Michaelis-Menten equation for the rate of enzymic processes, the reciprocal of velocity may be plotted on the y axis against the reciprocal of substrate concentration on the x axis. Determination of K_m and V_{max} is made possible by direct graphical measurement of both slope and x-axis intercept.

43. The answer is B. *(Mahler, ed 2. pp 295-297.)* In the enzymic reaction, increasing the proton concentration would decrease the V_{max} without affecting K_m. Therefore, the inhibition pattern would be noncompetitive, characterized in the double reciprocal (Lineweaver-Burk) plot by an alteration in both slope and intercept.

44. The answer is A. *(Mahler, ed 2. pp 295-299.)* Competitive inhibition kinetics are defined as effects that increase K_m of an enzyme without affecting V_{max}. Thus the double reciprocal plot changes in slope but not in y-axis intercept as enzyme concentration is altered.

45. The answer is B. *(Stryer, pp 119-120. White, ed 6. pp 211-212.)* Enzymes are catalysts and do not change the equilibrium of a chemical reaction. They accelerate both the forward and reverse reactions equally by lowering the energy of activation, without affecting the energy of reactants or products.

46. The answer is D. *(Stryer, pp 260-262.)* In the problem presented in the question, assuming that

$$K_{eq} = \frac{(C)}{(A)(B)} = 0.1,$$

then $\triangle G° = -2.3RT \log K_{eq} = -2.3RT \log (0.1) = 2.3RT$.

47. The answer is E. *(Stryer, p 127.)* In the question presented, the intercept of the double-reciprocal plot on the x axis is $-1/K_m$. Since $-1/K_m = -3$, $K_m = 0.33$. Similarly, from the observed y-axis intercept $(1/V_{max})$ the slope may be determined.

48. The answer is A. *(Stryer, pp 132-133.)* As shown on a double-reciprocal plot of enzyme activity rate vs. substrate concentration, a competitive inhibitor that increased K_m without affecting V_{max} would increase the slope without affecting the intercept on the y axis. On the graph presented in the question, such competitive inhibition would be represented by curve A, where curve X is the condition of no inhibition.

49. The answer is D. *(Stryer, p 127.)* The Michaelis-Menten hypothesis describes enzyme reaction kinetics. When such kinetics are represented by a double-reciprocal plot, the absolute value of the x-axis intercept gives a value of $1/K_m$, and the y-axis intercept, $1/V_{max}$.

50. The answer is E. *(White, ed 6. pp 267-268.)* Because free energy is a thermodynamic function of state, overall energy changes in biochemical reactions are independent of the reaction path. Systems at equilibrium have virtually no available free energy; only by utilizing free energy from without can such systems reach a state of disequilibrium.

51. The answer is C. *(White, ed 6. p 275.)* The redox system with the largest standard reduction potential among those listed in the question is cytochrome *a* (Fe^{3+}/Fe^{2+}). Electrons are passed from carriers with low standard reduction potentials to those with high standard reduction potentials.

52. The answer is D. *(Stryer, pp 258-260.)* The second law of thermodynamics is concerned with the concept of entropy, not free energy. Entropy is a measure of the degree of disorder in a system, and in a closed system any process goes spontaneously in the direction of increased entropy. The entropy of a system can decrease, as in the development of a highly structured biologic entity, provided the entropy of the surroundings increases sufficiently to give a positive value for the **sum** of the entropies of system and its surroundings.

53. The answer is B. *(Stryer, p 343.)* Atractyloside specifically blocks the facilitated exchange diffusion of ATP and ADP across the mitochondrial inner membrane, inhibiting both oxidative phosphorylation and electron transport. A specific carrier mediates the coupled flow of ATP and ADP across the inner mitochondrial membrane. The entry of ADP into the mitochondrial matrix is possible only with concomitant exit of ATP. This facilitated exchange diffusion is specifically blocked by atractyloside, which inhibits the carrier. The reactions of cytochrome a_3, cytochrome c, or coenzyme Q are not directly inhibited, but because atractyloside blocks oxidative phosphorylation without uncoupling it from electron transport, respiration ceases.

54. The answer is C. *(White, ed 6. p 365.)* Dinitrophenol, one of the uncouplers of oxidative phosphorylation, prevents the synthesis of ATP. Uncoupling involves a different mechanism from **inhibition** of oxidative phosphorylation, such as in the action of dicyclohexylcarbodiimide.

55. The answer is D. *(Davis, ed 2. p 127.)* Polymyxin, with one polar head and an aliphatic tail, resembles a cationic detergent and disrupts bacterial cell walls. It causes cell walls to become leaky, to lose intracellular constituents, and to admit dyes that normally are excluded.

56. The answer is A. *(Davis, ed 2. pp 112, 120.)* D-cycloserine is a structural analog of D-alanine; it blocks dipeptide synthetase to make D-alanyl-D-alanine. This competitive inhibition may be reversed by increasing the concentration of D-alanine.

57. The answer is E. *(White, ed 6. pp 225-227.)* For homotropic enzyme molecules, each substrate molecule bound causes a conformational change in the enzyme, enhancing subsequent substrate binding and reaction velocity.

58. The answer is C. *(Davis, ed 2. pp 127-128.)* Ionophores, which increase membrane permeability to specific ions, generate rings with a hydrophobic periphery that "dissolve" in the membrane. In addition, ionophores contain a hydrophilic interior having carboxyl groups that engage specific inorganic cations. Although antibiotics (e.g., valinomycin) that have these properties are bactericidal, they are insufficiently selective for clinical use.

59. The answer is B. *(Lenhinger, ed 2. pp 236, 283.)* A constitutive enzyme is one that is synthesized under all physiologic conditions and whose activity level is independent of induction or repression. For example, many nonregulated enzymes in a metabolic pathway are not induced but continuously synthesized at a level greater than the minimum required, in order that such enzymes will not be rate-limiting in the pathway. The terms allosteric and cooperative enzymes refer to enzymes that are regulated by small effector molecules, irrespective of whether the enzymes are inducible or constitutive. Any enzyme may, under given circumstances, be an inhibited enzyme.

60. The answer is D. *(Lehninger, ed 2. p 383.)* The term induction is applied to a case in which an enzyme's substrate leads to the *de novo* synthesis of the enzyme. Induction can be demonstrated with β-galactosidase in *E. coli*, wherein inducer and substrate functions are subject to separation and analysis.

61. The answer is B. *(Lehninger, ed 2. p 993.)* Most vertebrate organs are provided with a constant environment and do not need to respond to sudden changes. The liver, however, is exposed to varying rates of incoming nutrients and must respond to changes rapidly by induction or repression of its enzymes.

62. The answer is C. *(Mahler, ed 2. p 858.)* In the biosynthetic pathway outlined in the question, only a mutational block in enzyme C would produce an absence of Z while the pathway continued through M to L. Such a genetic block could arise from an absence of enzyme C, from its presence in marginal amounts, or from the presence of an abnormal C of greatly diminished activity.

63. The answer is A (1, 2, 3). *(Mahler, ed 2. pp 57-59.)* The partial double-bond character of peptide bonds prevents free rotation about such bonds. Peptide bonds usually have a *trans* configuration and are shorter than carbon-carbon single bonds.

64. The answer is E (all). *(Mahler, ed 2. pp 43-49.)* Isoleucine, methionine, proline, and phenylalanine all are amino acids with relatively nonpolar, hydrophobic side chains. The degree of hydrophobicity is directly proportional to the length of the side chain. Of this group, methionine has a unique capacity to bind metal ions by virtue of its sulfur atom.

65. The answer is A (1, 2, 3). *(Mahler, ed 2. pp 772-773. McGilvery, pp 662-663.)* Essential amino acids are those that must be supplied in the diet to assure normal growth, development, and health. While the dietary requirements for amino acids may be altered by various physiologic states, it is generally agreed that the amino acids essential to humans are leucine, isoleucine, valine, lysine, methionine, phenylalanine, tryptophan, threonine, and probably histidine.

66. The answer is A (1, 2, 3). *(McGilvery, pp 422-424, 496. Stryer, p 532.)* Heme is an iron-containing porphyrin. Glycine and succinyl CoA are precurors of 5-aminolevulinate from which porphyrins are derived. Creatine is formed by methylation of guanidinioacetate which is in turn derived from the transamination of glycine with arginine. Guanine is a purine whose ring structure is built upon the 5-phospho-α-D-ribosylpyrophosphate molecule by stepwise additions, which include the incorporation of a molecule of glycine to form carbons 4 and 5 and nitrogen 7 of the purine ring. There is no pathway for the direct incorporation of glycine into thymine, which is a pyrimidine derivative.

67. The answer is C (2, 4). *(McGilvery, pp 412-414, 424, 429, 662-663.)* Being essential amino acids, methionine and tryptophan by definition cannot be synthesized by mammalian tissues. Glycine, a nonessential amino acid, and choline can be synthesized, however, with serine as a precursor.

68. The answer is C (2, 4). *(Davis, ed 2. p 156.)* 6-Aminopenicillanic acid may be considered to be the condensation product of L-cysteine and D-valine. The 6-amino group receives a large variety of acyl groups, giving rise to a spectrum of penicillins.

69. The answer is D (4). *(Davis, ed 2. pp 120, 156.)* Cycloserine is essentially a closed alanine ring. In bacterial extracts, cycloserine inhibits the enzyme that racemizes L-alanine to yield D-alanine and the enzyme that incorporates D-alanine into dipeptides: it thus blocks two early steps in cell wall synthesis.

70. The answer is B (1, 3). *(Davis, ed 2. pp 154-155, 159, 298.)* Sulfisoxazole, a sulfonamide analog, and *p*-aminosalicylic acid inhibit folate metabolism. Isoniazid has a structural similarity to nicotinamide and to pyridoxamine. It inhibits conversion of nicotinamide to nicotinamide adenine dinucleotide phosphate (NAD) and enzyme reactions that require pyridoxal phosphate as a cofactor. It also inhibits synthesis of the cell wall constituent mycolic acid, but it is not clear which of these inhibitory effects mediates the antimicrobial action of isoniazid. Tetracycline causes a bacterostatic inhibition of protein synthesis.

71. The answer is A (1, 2, 3). *(Mahler, ed 2. pp 164-170.)* In general, it is the primary structure that determines the higher order structure of proteins. (There are certain well known exceptions, such as insulin.) Disulfide bonds are usually of more importance in stabilizing a protein's conformation than in determining it. While charged amino acids tend to be exposed to water, hydrophobic side chains tend to be buried within the center of the molecule away from water.

72. The answer is A (1, 2, 3). *(Mahler, ed 2. pp 129-133.)* Gamma immunoglobulins are composed of two heavy and two light chains with intra- and interchain disulfide linkages. There are two antigen binding sites per molecule. The N-terminus of both heavy and light chains is the variable region, whereas the C-terminus is constant. In multiple myeloma, large quantities of a homogeneous light chain (Bence Jones protein) may be excreted in the urine.

73. The answer is D (4). *(Mahler, ed 2. pp 762-769.)* Trypsinogen and chymotrypsinogen are synthesized in the pancreas. They are the zymogen precursors of the endopeptidases trypsin and chymotrypsin, respectively. Limited digestion with trypsin converts the zymogens into the active enzymes. The two zymogens have considerable homology in amino acid sequence.

74. The answer is E (all). *(Mahler, ed 2. pp 325-376.)* Most enzymes are proteins with a molecular weight in excess of 5,000 daltons. They have a high degree of specificity for substrate; they also catalyze a specific reaction by decreasing its activation energy. Most, but not all, enzymes have optimal activity in the vicinity of neutral or physiologic pH.

75. The answer is B (1, 3). *(Stryer, pp 735-738.)* In the question presented, an immunoglobulin fragment A, consisting of the light chain and part of the heavy chain, contains the antibody-combining site and the "hypervariable regions." Fragment A is known as Fab and fragment B as Fc.

76. The answer is E (all). *(Mahler, ed 2. pp 419-420.)* The various porphyrins are derived from a tetrapyrrole ring nucleus. They are able to form metal chelates with many ions including iron, copper, zinc, and cobalt. Protoporphyrin IX complexed with iron is referred to as heme and is found in hemoglobin, myoglobin, cytochrome, and catalase.

77. The answer is E (all). *(Harper, ed 16. p 531.)* Hemoglobin, myoglobin, and cytochrome *c* all contain iron in a heme ring. Peroxidase also contains iron, but without the heme moiety. Catalase, another enzyme, also contains iron in the ferric state which, unlike that in peroxidase, is reduced to the ferrous form only with difficulty.

78. The answer is A (1, 2, 3). *(White, ed 6. pp 941-945.)* The complement system, a self-assembling system consisting of 11 serum globulin proteins, mediates some of the antigen-antibody reaction effects upon cells. Cell lysis can be brought about by an initial reaction of antibody with antigen on the cell surface. The complement system acts with antibodies to produce cell lysis. The 11 proteins have been grouped into three functional units; a recognition unit (three proteins), an activation unit (three proteins), and a membrane attack unit (five proteins).

79. The answer is E (all). *(Mahler, ed 2. pp 303-309. White, ed 6. pp 220-222.)* The Monod model of allosteric enzymes holds that such enzymes consist of polymers of subunits each containing both an active catalytic site and an allosteric site for the binding of modifiers. It further assumes that conformational changes occur in a concerted fashion; for instance, all subunits in a given polymer are in the same state at the same time. In a K-system, the different conformational states of the subunits differ with respect to their affinities for various ligands, such as substrates, inhibitors, or activators.

80. The answer is D (4). *(Stryer, pp 333-335.)* The redox potential of a couple such as fumarate/succinate (or Fe^{3+}/Fe^{2+}) is a measure of electron transfer potential. The reduction (gain of electrons) of fumarate to succinate has a redox potential of +0.03 volts relative to the standard H_2/H^+ couple, while that for the reduction of ferric sulfate (Fe^{3+}) to ferrous sulfate (Fe^{2+}) is +0.77 volts. Thus ferric iron has a greater affinity for electrons than fumarate and, in the mixture of the two couples, succinate will be oxidized (loss of electrons) to fumarate and ferric sulfate reduced (gain of electrons) to ferrous sulfate.

81. The answer is B (1, 3). *(White, ed 6. pp 223-224.)* When a modifier binds at the allosteric site, it affects the active site by altering V_{max} and K_m. Binding of both substrate and modifier are, of course, concentration-dependent. The velocity of an allosteric enzyme reaction depends on the concentrations of both the substrate and the modifier.

82. The answer is C (2, 4). *(White, ed 6. p 227.)* Examples of allostery may include enzymes like *E. coli* aspartate transcarbamoylase, or binding proteins like hemoglobin. Nonregulatory enzymes, such as phosphoglucose isomerase and lactate dehydrogenase, do not exhibit sigmoidal kinetics.

83. The answer is C (2, 4). *(Mahler, ed 2. pp 666-671.)* Hemoglobin is a tetrameric hemoprotein whose oxygen saturation curves exhibit sigmodal kinetics because of cooperative interactions among the four binding sites. Oxygen is bound to hemoglobin without changing the redox state of the iron from the ferrous state. Carbon monoxide and cyanide both bind to hemoglobin more tightly than does oxygen itself.

84. The answer is C (2, 4). *(White, ed 6. pp 357-369.)* Two hypotheses that attempt to explain the mechanism of oxidative phosphorylation are the chemiosmotic hypothesis of Mitchell and the chemical coupling hypothesis. The Pasteur effect is the inhibition of glycolysis by oxygen, which Warburg's hypothesis related to the metabolism of tumors.

85. The answer is B (1, 3). *(Mahler, ed 2. pp 680-684.)* The standard redox potential of cytochrome *b* is lower than that of either cytochrome *c* or *a*. Cytochrome *b* does not interact directly with oxygen, cyanide, or carbon monoxide. It reacts with cytochrome *a* through cytochrome *c*.

86. The answer is B (1, 3). *(Mahler, ed 2. pp 27-32.)* The oxidation-reduction potential is a quantitative measure of the tendency of a compound to donate or accept electrons. The hydrogen electrode is arbitrarily assigned a standard potential of zero. Compounds with negative reduction potentials will tend to reduce protons to hydrogen gas, while compounds with positive reduction potentials will tend to accept electrons from H_2. Standard redox potentials apply to the situations wherein all reactants and products are present at concentrations of 1 M or tensions of 1 atm; it may be necessary to correct for actual concentrations. The standard free energy change of a reaction can be calculated from the standard reduction potential (E_o) by the formula $\Delta G° = -nF \Delta E_o$.

87. The answer is C (2, 4). *(Mahler, ed 2. pp 690-692.)* There are thought to be three sites of energy conservation in the mitochondrial electron transport chain that leads to the production of ATP: between NAD and a flavoprotein (site I); between cytochrome *b* and cytochrome *c* (site II); and between cytochrome *c* and cytochromes a, a_3 (site III).

88. The answer is A (1, 2, 3). *(White, ed 6. pp 975-976.)* Sickle cell anemia, which results from a single amino acid substitution in the hemoglobin beta-chain, occurs in varying degrees in persons with S-S homozygous genetic constitution. Individuals with a single allele for hemoglobin S (heterozygotes) have sickle cell trait; though they may have a mild microcytic anemia, they do not have sickle cell disease. The heterozygous state confers protection against malarial parasites.

89. The answer is E (all). *(Davis, ed 2. pp 135-136, 163, 357.)* Synthesis of the two bacterial enzymes and the transport protein listed in the question all are inducible. Immunoglobulin synthesis is inducible by exposure to antigen to which the cell has been sensitized; thus, antibodies are categorically inducible.

90. The answer is B (1, 3). *(White, ed 6. p 378.)* The polymeric structure of some enzymes may result in the presence of related but different forms of the same enzyme in tissue. Separately encoded monomeric subunits are assembled in different proportions into distinct multimeric complexes having identical substrate and cofactor specificities but usually exhibiting different K_m values for either, or both. Isozymes are detected by differences in electrophoretic mobility.

91. The answer is C (2, 4). *(Harper, ed 16. pp 560-561.)* Heparin is a naturally occurring mucopolysaccharide that prevents the clotting of blood by interfering with a number of steps in the coagulation cascade. Plasmin is a circulating enzyme that hydrolyzes fibrin clots to form soluble products. Fibrinogen is the substrate acted upon by thrombin to yield the fibrin mesh of blood clots, and platelets are the agents first activated at the site of disruption in blood vessels to prevent hemorrhage.

92. The answer is E (all). *(Davis, ed 2. pp 131-132.)* All the characteristics listed in the question apply to the galactoside transport system of *E. coli*. While energy is required for active transport, **facilitated** transport of β-galactosides does not require the expenditure of energy. This type of transport characterizes active membrane transport systems that experience "downhill flow," becoming detached from the source of energy.

93. The answer is D (4). *(Davis, ed 2. pp 409-411.)* Heavy and light chains of IgG may be separated by reduction of interchain disulfide bonds with mercaptoethanol. Papain digests IgG into three proteolytic fractions (Fc and two Fab portions), while pepsin treatment yields almost indistinguishable $(Fab')_2$ fragments and digests completely the Fc fragment.

94-95. The answers are: 94-C, 95-A. *(Mahler, ed 2. pp 775-789.)* Histidine is synthesized (in microorganisms) from ribose 5-phosphate as an early precursor. Proline is synthesized in animal tissues, as well as in microorganisms, from glutamate as a precursor. The classification in animals of "essential" and "nonessential" amino acids reflects animals' incapacity, compared to plants and microorganisms, to synthesize all the amino acids they require.

96-99. The answers are: 96-D, 97-B, 98-E, 99-C. *(Mahler, ed 2. pp 43-49, 139.)* α-Amino acids, the monomeric units that result from complete protein hydrolysis, are conveniently classified according to the properties of their side chains. Thus, cysteine contains sulfur, and lysine is a basic amino acid. The side chain of leucine is an aliphatic, branched chain. Phenylalanine has an aromatic side chain. Protein hydrolysates contain, in addition to α-amino acids, the two α-imino acids proline and hydroxyproline, which differ from other amino acids in having amino and carboxyl groups that do not share the same carbon atom.

100-102. The answers are: 100-C, 101-D, 102-B. *(Mahler, ed 2. pp 478-479. White, ed 6. pp 720-721, 1135-1136.)* Hair and nails are composed of keratin which contains a large amount of the disulfide amino acid cystine (approximately 14 percent in human hair). The enzyme tyrosinase catalyzes a series of reactions leading from tyrosine to dihydroxyphenylalanine and to melanin, which is a skin and hair pigment. Twenty percent of the amino acids in collagen, a triple-stranded protein formed by fibroblasts, is proline and hydroxyproline.

Chondroitin sulfate, a polysaccharide, is composed of either glucuronic or iduronic acid, according to whether the sulfate is in the A and C form or the B form, respectively.

103-104. The answers are **103-D, 104-A**. *(Mahler, ed 2. p 115. Stryer, p 26.)* Cleavage of protein disulfide bridges can be achieved by reaction with mercaptoethanol. The peptide bond formed between the α-carboxyl group of methionine and the α-amino group of the next amino acid residue is split during reaction of the protein with cyanogen bromide. The proteolytic enzymes, trypsin and chymotrypsin, hydrolyze peptide bonds on the carboxyl side of arginine and lysine residues, and on the carboxyl side of aromatic and other nonpolar residues, respectively. 2,4-Dinitrophenol (DNP) does not cleave peptide bonds, but a derivative, fluorodinitrobenzene, reacts with terminal amino acids in proteins (Sanger reagent).

105-107. The answers are: **105-D, 106-C, 107-E**. *(Mahler, ed 2. pp 116-117.)* Trypsin, the most specific proteolytic enzyme, is an endopeptidase that preferentially cleaves after basic amino acids (lysine or arginine). Chymotrypsin is an endopeptidase that preferentially cleaves after aromatic amino acids (tryptophan, tyrosine). These specific cleavage properties make trypsin and chymotrypsin particularly useful in determining the sequence of amino acids in long polypeptides. Carboxypeptidase is an exopeptidase that cleaves at the carboxyl terminus of a polypeptide.

108-110. The answers are: **108-A, 109-C, 110-D**. *(Stryer, pp 124-126.)* Differential equations for the kinetics of enzyme-catalyzed reactions involve several parameters defined by the rate constants of such reactions. For the reaction $S + E \underset{k_2}{\overset{k_1}{\rightleftharpoons}} S \cdot E$, the dissociation constant K_s equals the Michaelis constant, inasmuch as k_3 is absent and the expression $\frac{k_2 + k_3}{k_1}$ becomes k_2/k_1. Where the enzyme-substrate complex proceeds irreversibly at rate k_3 to yield product P in the equation reproduced in the question, the reaction velocity, which is proportional to enzyme-substrate concentration ES, is expressed as $v = \frac{k_3 \, ES}{Km + S}$. Maximal velocity ($V_{max}$) obtains at initial enzyme concentration (E_T), and for a constant substrate concentration is defined as $V = k_3 E_T$.

Nucleic Acids and Molecular Genetics

DIRECTIONS: Each question below contains five suggested answers. Choose the **one best** response to each question.

111. Only DNA, and not RNA, would be radioactively labeled in an animal fed tritiated

(A) adenine
(B) cytosine
(C) guanine
(D) thymine
(E) uracil

112. Which of the following nucleic acid bases is found in mRNA but not in DNA?

(A) Adenine
(B) Cytosine
(C) Guanine
(D) Uracil
(E) Thymine

113. Unusual nucleotide bases are found primarily in

(A) rRNA
(B) mRNA
(C) tRNA
(D) nucleolar DNA
(E) mitochondrial DNA

114. A severe hereditary form of gout results from the absence of an enzyme involved in the "salvage pathways" of nucleotide metabolism. This enzyme is

(A) hypoxanthine-guanine phos-phoribosyltransferase
(B) aspartate transcarbamoylase
(C) thymidylate kinase
(D) adenylate deaminase
(E) nucleoside phosphorylase

115. Which of the following compounds does NOT contain a methyl group donated from S-adenosyl methionine?

(A) Creatine phosphate
(B) Epinephrine
(C) Melatonin
(D) Phosphatidylcholine
(E) Thymine

116. <u>S-Adenosylmethionine</u> is shown below with five substituent groups labeled A through E. Which group is S-adenosylmethionine able to donate in creatine synthesis?

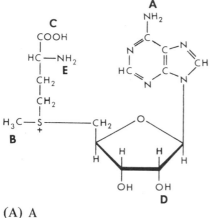

(A) A
(B) B
(C) C
(D) D
(E) E

117. The <u>end product of purine</u> metabolism that is excreted by humans is

(A) allantoic acid
(B) orotic acid
(C) urea
(D) uric acid
(E) xanthine

118. Which of the following is a metabolic pathway <u>common to</u> <u>bacteria and humans?</u>

(A) Purine synthesis
(B) Nitrogen fixation
(C) Cell wall mucopeptide synthesis
(D) Noncyclic photophosphorylation
(E) Fermentation to ethyl alcohol

119. Carbons 4 and 5 in the purine nucleus shown below are obtained from

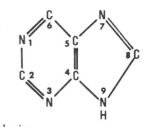

(A) glycine
(B) alanine
(C) acetate
(D) aspartate
(E) glutamate

120. Which of the following is <u>true</u> of <i>de novo</i> <u>pyrimidine synthesis</u> but not of purine biosynthesis?

(A) The bases are synthesized while attached to ribose 5-phosphate
(B) One-carbon fragments are donated by folic acid derivatives
(C) Carbamoyl phosphate donates a carbamoyl group
(D) Glycine is incorporated intact
(E) Glutamine is a nitrogen donor

121. It is true of <u>mitochondrial—</u> but not of cytoplasmic—<u>carbamoyl</u> <u>phosphate synthetase</u> that it

(A) is inhibited by uridine triphosphate
(B) is activated by acetylglutamate
(C) is involved in pyrimidine biosynthesis
(D) is present in relatively low activity
(E) uses glutamine as a nitrogen source

122. Which of the following statements about the lettered ring constituents of uracil shown below is true?

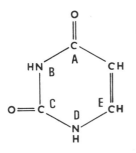

(A) The carbon atom lettered A is derived from the β-carboxyl group of aspartic acid
(B) The nitrogen atom lettered B is derived from the ε-amino group of lysine
(C) The carbon atom lettered C is derived from the carboxyl of methionine
(D) The nitrogen atom lettered D is derived from the amino group of glutamine
(E) The carbon atom lettered E is donated by folic acid as a free methyl group

123. In *E. coli*, pyrimidine feedback inhibition controls the activity of

(A) dihydro-orotase
(B) orotidylic pyrophosphorylase
(C) reductase
(D) aspartate transcarbamoylase
(E) hydroxymethyl cytidylate synthetase

124. The degradation of RNA by pancreatic ribonuclease produces

(A) nucleoside 2'-phosphates
(B) nucleoside 5'-phosphates
(C) nucleoside 3'-phosphates only
(D) nucleoside 3'-phosphates and oligonucleotides
(E) oligonucleosides only

125. The sedimentation coefficient of a DNA molecule depends on all the following physical characteristics EXCEPT

(A) partial specific volume
(B) diffusion coefficient
(C) optical density
(D) molecular weight
(E) molecular shape

126. The point called T_m (melting temperature) for double-stranded DNA is represented by which of the following letters in the diagram below?

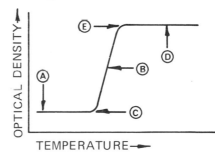

(A) A
(B) B
(C) C
(D) D
(E) E

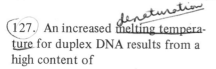

127. An increased melting tempera-ture for duplex DNA results from a high content of

(A) adenine (A) + guanine (G)
(B) cytosine (C) + thymine (T)
(C) adenine + thymine
(D) cytosine + guanine
(E) adenine + cytosine

128. The Watson-Crick model of DNA structure shows

(A) a triple-stranded structure
(B) the DNA strands running in opposite directions
(C) pair-bonding between bases A and G
(D) covalent-bonding between bases
(E) the phosphate backbone to be on the inside of the DNA helix

129. All the following statements concerning histones are true EXCEPT that they

(A) are proteins of relatively low molecular weight
(B) are rich in arginine and lysine
(C) are noncovalently attached to DNA in stoichiometric amounts
(D) exhibit known interspecies homologies in their amino acid sequences
(E) exhibit a wide variety of non-homologous molecular structure

130. During DNA replication, the sequence 5'-TpApGpAp-3' would produce which of the following complementary structures?

(A) 5'-TpCpTpAp-3'
(B) 5'-ApTpCpTp-3'
(C) 5'-UpCpUpAp-3'
(D) 5'-GpCpGpAp-3'
(E) 3'-TpCpTpAp-3'

131. AUG, the only identified codon for methionine, is important as

(A) the site of attachment for the 30-S ribosomal particle
(B) the recognition site on the trans-fer RNA
(C) a releasing factor for peptide chains
(D) a chain-terminating codon
(E) a chain-initiating codon

132. The fact that DNA bears the genetic information of an organism implies that

(A) base composition should be identical from species to species
(B) viral infection is accomplished by transfer of protein into the host cell
(C) DNA from different tissues in the same organism should usually have the same base composition
(D) DNA base composition should change with age and nutritional state of an organism
(E) DNA occurs in small ring-shaped structures

133. DNA replicates in a semiconservative manner. If a completely radioactive double-stranded DNA molecule undergoes two rounds of replication in a solution free of radioactive label, what is the radioactivity status of the resulting four DNA molecules?

(A) Half should contain no radioactivity
(B) All should contain radioactivity
(C) Half should contain radioactivity in both strands
(D) One should contain radioactivity in both strands
(E) None should contain radioactivity

134. Which of the following mutations is most likely to be lethal?

(A) Substitution of adenine for cytosine
(B) Substitution of cytosine for guanine
(C) Substitution of methylcytosine for cytosine
(D) Deletion of three nucleotides
(E) Insertion of one nucleotide

135. The anticodon in tRNA that corresponds to the codon ACG in mRNA is

(A) UGC
(B) TGC
(C) GCA
(D) CGU
(E) CGT

136. Thymine dimers that occur in DNA as a mutational event

(A) do not stop replication
(B) are repaired by an enzyme system that includes the enzyme ligase
(C) are read as frame-shift mutations
(D) are catalyzed by the enzyme thymine dimerase
(E) create covalent bonds between thymines on opposite nucleotide strands

137. Sickle cell anemia is the clinical manifestation of homozygous genes for an abnormal hemoglobin molecule. The mutational event responsible for the mutation in the beta chain is

(A) crossing over
(B) insertion
(C) deletion
(D) nondisjunction
(E) point mutation

138. The amino terminus of all polypeptide chains at the time of synthesis in *E. coli* is thought to be which of the following amino acid residues?

(A) Methionine
(B) Serine
(C) *N*-Formylmethionine
(D) *N*-Formylserine
(E) Glutamate

139. Wobble is best described as

(A) the ability of certain anticodons to pair with codons that differ at the third base
(B) a mechanism that allows for peptide extension in the 50-S subunit of the ribosome
(C) an error in translation induced by streptomycin
(D) the induction of lysogenic phages to become virulent
(E) thermal motions leading to local denaturation of the DNA double helix

140. How many different codons are capable of terminating polypeptide chain elongation in protein synthesis?

(A) One
(B) Two
(C) Three
(D) Four
(E) Five

141. All the following statements about DNA-primed RNA synthesis are true EXCEPT that

(A) RNA polymerase catalyzes the formation of phosphodiester bonds only in the presence of DNA
(B) RNA polymerase requires a primer in the transcription process
(C) the direction of growth of the RNA chain is from 5'- to 3'-end
(D) only one strand of DNA serves as template in most circumstances
(E) the RNA chain synthesized is never circular

142. Sigma factor is best described as a

(A) subunit of RNA polymerase responsible for the specificity of the initiation of transcription of RNA from DNA
(B) subunit of DNA polymerase that allows for bidirectional synthesis in both 5'- to-3' and 3'- to-5' directions
(C) subunit of the 50-S ribosome that catalyzes peptide bond synthesis
(D) subunit of the 30-S ribosome to which mRNA binds
(E) factor that forms the bridge between the 30-S and 50-S particles constituting the 70-S ribosome

143. The first step in the incorporation of amino acids into polypeptide chains in prokaryote cells requires

(A) binding of formylmethionyl tRNA to the ribosome
(B) binding of the 30-S to the 50-S ribosomal subunit
(C) binding of the mRNA to the 30-S ribosomal subunit
(D) participation of aminoacyl-tRNA synthetase
(E) initiation factors

144. Which of the following reactions requires guanosine triphosphate?

(A) The synthesis of peptides at ribosomal sites
(B) The release of peptides from polyribosomes
(C) The activation of amino acids by aminoacyl-RNA synthetase
(D) The activation of arginine by arginyl-RNA synthetase
(E) The formation of polyribosomal units

145. Puromycin has been useful in studying protein synthesis because it

(A) alters the binding of tRNA to mRNA
(B) binds to the 30-S subunit of the ribosome
(C) releases peptide material from ribosomal complexes
(D) inhibits the amino acid activating enzyme
(E) prevents the formation of polysomes

146. Tetracycline prevents synthesis of polypeptides by

(A) competing with mRNA for ribosomal binding sites
(B) blocking mRNA formation from DNA
(C) releasing peptides from mRNA-tRNA complexes
(D) inhibiting tRNA binding to mRNA
(E) preventing amino acid binding by tRNA

147. In prokaryotes, chloramphenicol has which of the following biochemical effects?

(A) Causes premature release of the polypeptide chain
(B) Causes misreading of the mRNA
(C) Depolymerizes DNA
(D) Inhibits protein synthetic activities of the 30-S ribosomal subunit
(E) Inhibits protein synthetic activities of the 50-S ribosomal subunit

148. 5-Bromouracil is mutagenic because

(A) it inserts extra base pairs during replication
(B) the bromine does not fit into the double helix as does the methyl group of thymine
(C) when it is incorporated into DNA in place of thymine, guanine may be paired with it in the next cycle of replication
(D) when it is incorporated into RNA in place of adenine, codons are misread
(E) it deaminates adenine, guanine, or cytosine

149. Translation results in a product known as

(A) protein
(B) tRNA
(C) mRNA
(D) rRNA
(E) DNA

150. Which of the following statements about ribosomes is true?

(A) They are an integral part of
transcription
(B) They are found both free in the
cytoplasm and bound to membranes
(C) They are bound together so
tightly they cannot dissociate
under physiologic conditions
(D) They are composed of RNA,
DNA, and protein
(E) They are composed of three
subunits of unequal size

151. DNA that has been subject to
thermal denaturation is characterized by

(A) cleavage of the phosphodiester
links between bases
(B) the formation of a triple helix
(C) a broad ($10°C$) denaturation
range for a homogenous DNA
(D) a melting-point temperature
that varies directly with guanine-
cytosine base pair content
(E) a decrease in light absorption
at 260 nm

152. Peptide chain elongation involves all the following EXCEPT

(A) peptidyl transferase
(B) GTP
(C) Tu, Ts, and G factors
(D) formylmet-tRNA
(E) mRNA

153. The cell wall of gram-positive
bacteria is

(A) not weakened by lysozyme
(B) less than 15 Å thick
(C) composed partly of N-acetyl-
muramic acid
(D) characterized by a lipoprotein
layer
(E) composed of only L-amino
acids

154. In the process known as transformation, bacteria

(A) are transformed by a bacterial
phage
(B) pass their DNA from cell to
cell
(C) take up pieces of DNA from
the media
(D) develop point mutations
(E) develop frame-shift mutations

155. The classic phage shown to
cause specialized transduction in *E.
coli* is

(A) ϕX 10
(B) T 4
(C) phage λ
(D) K 12
(E) P 22

156. The transfer of a fragment of
donor chromosome to a recipient cell
by a bacteriophage that has been produced in the donor cell is called

(A) transduction
(B) transformation
(C) conjugation
(D) integration
(E) mutation

157. The process of transferring a large piece of chromosomal material between donor and recipient microbial cells is known as

(A) generalized transduction
(B) restricted transduction
(C) transformation
(●) conjugation
(E) repression transfer

158. Which of the following statements concerning RNA and DNA polymerases is true?

(A) RNA polymerases use nucleoside diphosphates, rather than nucleoside triphosphates, to form polynucleotide chains
(B) RNA polymerases require primers and add bases at the $5'$-end of the growing polynucleotide chain
(C) DNA polymerases can add nucleotides at both ends of the chain
(D) DNA polymerases can only synthesize DNA by copying information from an RNA template
(●) All RNA and DNA polymerases can only add nucleotides at the $3'$-end of the growing polynucleotide chain

159. Which of the following statements about interferon is true?

(A) It is virus-specific
(B) It is a bacterial product
(C) It is a synthetic antiviral agent
(D) It inhibits viral multiplication in all cells
(●) It requires expression of cellular genes

160. The most likely cause of a spontaneous point mutation in an *E. coli* culture is

(●) a tautomeric shift of a hydrogen atom
(B) a break in the sugar-phosphate backbone of DNA
(C) the insertion of a single base pair
(D) crosslinking of strands
(E) mutarotation of deoxyribose

161. A frame-shift mutation caused by the insertion or deletion of a base pair is most likely to be caused by

(●) acridine derivatives
(B) 5-bromouracil
(C) azaserine
(D) ethyl ethanesulfonate
(E) azathioprine

162. In a patient who has phenylketonuria, a hereditary disorder of metabolism, which of the following statements is true?

(A) Phenylalanine and tyrosine both are nonessential amino acids
(●) Phenylalanine and tyrosine both are essential amino acids
(C) Phenylalanine is an essential amino acid, tyrosine is not
(D) Tyrosine is an essential amino acid, phenylalanine is not
(E) Phenylalanine can be replaced by tyrosine in the diet

163. The location on the lactose operon shown below where RNA polymerase and sigma factor bind initially before initiating transcription is

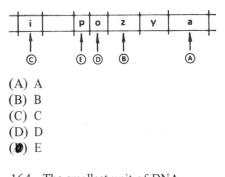

(A) A
(B) B
(C) C
(D) D
(✏) E

164. The smallest unit of DNA capable of coding for the synthesis of a polypeptide is the

(A) operon
(✏) cistron
(C) promoter
(D) replicon
(E) repressor gene

165. In a Hfr (high frequency of recombination) x F⁻ cross, the initiation point of chromosome transfer is determined by

(A) the genotype of the recipient F⁻ strain
(B) the genotype of the Hfr strain
(C) the phenotype of the Hfr strain
(D) the conditions of the conjugation
(✏) the integration of the F factor

166. When the distance between genetic loci is sufficiently long to permit double crossovers, the recombination frequency between the two genetic loci

(✏) is always less than the sum of intervening frequencies
(B) is always more than the sum of intervening frequencies
(C) is equal to the sum of intervening frequencies
(D) exhibits a direct linear relationship with the crossover frequency
(E) cannot be qualitatively related to the sum of intervening frequencies

167. The percentage of a bacterial chromosome usually involved in transformations is

(✏) 1 percent
(B) 10 percent
(C) 25 percent
(D) 50 percent
(E) 100 percent

168. A common characteristic of conjugating bacteria is

(A) the transfer of the bacterial chromosome
(B) the formation of recombinants
(C) the conversion of the female cells to male cells
(✏) the presence of F pili
(E) the presence of a high-frequency recombinant

169. The electron micrograph below shows a circular, double-stranded molecule of DNA having a molecular weight of ten million daltons. It is typical of DNA found in

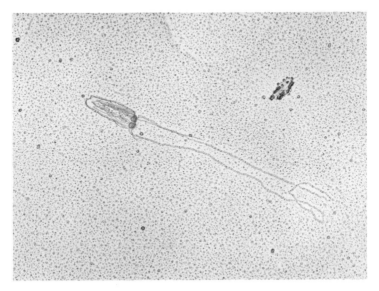

(A) the nuclei of eukaryotes
(B) mammalian chromosomes
(C) *E. coli*
(D) mitochondria
(E) lysosomes

---

170. In the cross of *E. coli* strain strsF′lac⁺ with strain lac⁻strʳ, the desired recipients may be selectively grown on

(A) lactose-minimal plates
(B) lactose-minimal streptomycin plates
(C) glucose-minimal streptomycin plates
(D) glucose-minimal plates
(E) arabinose-minimal plates

171. A potent inhibitor of protein synthesis that acts as an analog of aminoacyl-tRNA is

(A) mitomycin C
(B) streptomycin
(C) nalidixic acid
(D) rifampicin
(E) puromycin

172. The non-overlapping nature of the genetic code is supported by the fact that

(A) poly(U-G) directs the synthesis of poly(cys-val)
(B) a single base mutation alters only one amino acid in the resulting protein
(C) most amino acids are coded by more than one triplet
(D) ribosomal bindings of amino-acyl-tRNA is stimulated by tri-nucleotides
(E) several terminator triplets have been discovered

173. The major implication of the *cis/trans* test for a gene in hetero-zygous diploid cells is that when two strains have mutant sites lying in

(A) different genes, they show com-plementation to give a wild phenotype in only the *cis* arrange-ment
(B) different genes, they show com-plementation to give a wild phenotype in only the *trans* arrangement
(C) the same gene, they show com-plementation to give a wild phenotype in only the *cis* arrange-ment
(D) the same gene, they show com-plementation to give a wild phenotype in only the *trans* arrangement
(E) the same gene, they show com-plementation to give a wild phenotype in either the *cis* or *trans* arrangement

174. Xeroderma pigmentosum is an inherited human skin disease that causes a variety of phenotypic changes in skin cells exposed to sun-light. The molecular basis of the disease appears to be

(A) a rapid water loss caused by defects in the cell membrane permeability
(B) the inactivation of temperature-sensitive transport enzymes in sunlight
(C) the induction of a virulent pro-virus upon ultraviolet exposure
(D) the inability of the cells to synthesize carotenoid-type compounds
(E) a defect in an excision-repair system that removes thymine dimers from DNA

175. F' factors frequently are lost from their bacterial hosts as a result of storage. An F'lac⁺ factor is best maintained in

(A) a lac⁻recA⁻ strain
(B) a lac⁺recA⁺ strain
(C) a lac⁺recA⁻ strain
(D) a lac⁻recA⁺ strain
(E) an Hfr strain

176. A system of contiguous structural genes and controlling elements for the synthesis of various enzymes in a metabolic pathway is best described as

(A) a genome
(B) a muton
(C) a cistron
(D) a codon
(E) an operon

177. How many high-energy phosphate bonds are expended in a complete cycle in protein synthesis, from a free amino acid to its incorporation into a peptide?

(A) 0
(B) 1
(C) 2
(D) 3
(E) 4

178. In studies of the mechanism of bacterial DNA replication, 5-bromouracil often is used as an analog of thymidine in order to

(A) cause specific frame-shift mutations for sequencing studies
(B) stop DNA synthesis at sites of thymidine incorporation
(C) provide a reactive group in the DNA for the preparation of DNA affinity supports
(D) synthesize a denser DNA that can be identified by centrifugation
(E) create specific sites in the DNA for mild chemical cleavage

179. The point of integration of the prophage in lysogenic bacteria may be determined by conjugation with a phage-sensitive recipient. The phenomenon is called

(A) gradient of transmission
(B) mapping by interrupted matings
(C) episome transfer
(D) phenotypic lag
(E) zygotic induction

180. Acridines, such as proflavine, induce which of the following classes of genetic mutation?

(A) Chromosomal rearrangement
(B) Frame-shift
(C) Transversion
(D) Transition
(E) Specific transition G:C A:T

181. The preferential transcription and translation of the T4 phage genome by the host cell is best explained by

(A) viral-specific changes in the host RNA polymerase which increase the specificity of phage DNA transcription
(B) viral-specific changes in the host ribosomes which increase the specificity of phage mRNA translation
(C) the synthesis of viral-coded nucleases specific for host DNA
(D) the synthesis of viral-coded proteases specific for host proteins
(E) repression of host DNA by viral regulators

182. Lysogenic bacteria are immune to superinfection because

(A) the prophage may be induced to enter the lytic cycle by excision from the bacterial chromosome
(B) the prophage is not inherited as part of the bacterial chromosome
(●) lysogeny and immunity result from the operation of the same control system that represses phage genes needed for lytic development
(D) bacterial phage receptors in the cell wall are inactivated by the process of lysogeny
(E) they actively synthesize phage-specific nucleases that degrade any phage DNA introduced into the bacteria

183. The *lac* operon is a control system which, in wild type *E. coli*, is

(A) repressible, with negative control
(●) inducible, with negative control
(C) repressible, with positive control
(D) inducible, with positive control
(E) constitutive, with control by attenuation

184. Positive control of induction is best described as a control system in which an operon functions

(A) unless it is switched off by a derepressed repressor protein
(B) unless a repressor protein, which is activated by a co-repressor, switches it off
(C) only after a repressor protein is inactivated by an inducer
(D) only after an inducer protein, which can be inactivated by a co-repressor, switches it on
(●) only after an inducer protein, which is activated by an inducer, switches it on

185. The function of a repressor protein in an operon system is to prevent protein synthesis by binding to

(A) an initiator formylmethionine (fmet)-tRNA
(B) the ribosome
(C) the RNA polymerase, blocking transcription
(D) a specific region of the mRNA, preventing translation to protein
(●) a specific region of the operon, preventing transcription of structural genes

186. Hot spots are best described as

(A) sites in proteins of high suscep-
tibility to amino acid modifica-
tion

(B) localized areas of energy build-
up within the cell

(C) DNA regions of high recombination frequency

(D) sites in DNA of high suscep-
tibility to mutagenesis

(E) radioactive fragments of DNA
formed during in vitro
replication experiments

187. In the method of genetic transfer called transduction,

(A) the donor bacterium unites with
a recipient bacterium and transfers to the recipient a part of its
chromosome

(B) purified DNA from the cells of
a donor strain is used to induce
permanent hereditary changes
in the cells of a second strain

(C) modified particles of bacterial
virus of low virulence act as vectors
of bacterial genes

(D) the sex factor of an F^+ strain is
transferred to an F^- strain

(E) large chromosomal regions are
transferred, thcrby facilitating
an overall mapping of genes

DIRECTIONS: Each question below contains four suggested answers of which **one** or **more** is correct. Choose the answer:

A	if	**1, 2, and 3**	are correct
B	if	**1 and 3**	are correct
C	if	**2 and 4**	are correct
D	if	**4**	is correct
E	if	**1, 2, 3, and 4**	are correct

188. Protein biosynthesis requires

(1) mRNA
(2) peptidyl transferase
(3) ribosomes
E (4) ATP

189. Covalent bonds in DNA include

(1) adenine → β-glycosidic linkage to C-1' of deoxyribose
B (2) phosphate → 2'-OH of deoxyribose
(3) phosphate → 5'-OH of deoxyribose
(4) phosphate → phosphate

190. Inhibitors of nucleic acid biosynthesis include

(1) rifampicin
B (2) chloramphenicol
(3) actinomycin D
(4) tetracycline

191. Donor cells of chromosomal DNA in an F^+ culture are called Hfr (high-frequency recombination) cells. Which of the following statements concerning them are true?

(1) Conjugation with an Hfr strain usually results in transfer of a complete chromosome
(2) Hfr cells transmit free F (sex factor) particles
(3) Hfr cells cannot revert to F^+ when the sex factor is detached from the chromosome
D (4) Each Hfr cell has its own specific origin and order of gene transfer

192. Which of the following statements about conjugation are true?

(1) Physical contact between donor and recipient cells is required
B (2) Conjugation results in the transfer of only small fragments of a donor's chromosomes
(3) Cells lacking the sex factor (F^-) can act only as recipients
(4) Cells possessing the sex factor (F^+) can act only as donors

193. Which of the following statements concerning the genetic mechanisms of bacterial drug resistance are true?

   (1) Mutations occur at a rate of 10^{-5} to 10^{-9} per cell, per generation
   (2) Recombination between two cells, each resistant to a different drug, can produce a cell resistant to both
   (3) Staphylococci that produce penicillinase do not arise by chromosomal gene mutation
   (4) A plasmid-carrying cell introduced into a cell population usually infects every other cell

194. The "early proteins" of the T-even bacteriophages include

   (1) enzymes for hydroxymethyl-cytosine production
   (2) head and tail subunits
   (3) DNA polymerase
   (4) lysozyme for cell wall degradation

195. The sex factor in an Hfr cell is

   (1) integrated in the chromosome
   (2) associated with a particular sequence of gene transfer
   (3) an endogenote
   (4) autotransferable in the same manner as the F factor

196. Inhibition of bacterial cell wall synthesis is the primary metabolic effect of

   (1) chloramphenicol
   (2) bacitracin
   (3) sulfonamide
   (4) penicillin

197. Microbial cells possessing the sex factor (F) can

   (1) act as males
   (2) act as females
   (3) conjugate with females
   (4) conjugate with males

198. Chloramphenicol has the ability to

   (1) interfere with cell wall synthesis in bacteria
   (2) act bacteriostatically against many microorganisms
   (3) interfere with RNA synthesis
   (4) interfere with the incoporation of amino acids

199. Genetic information coding for ribosomal RNAs is found in

   (1) bacteria
   (2) mitochondria
   (3) rickettsiae
   (4) viruses

200. Which of the following compounds exert their antibiotic effect on bacterial ribosomes?

   (1) Chloramphenicol
   (2) Gentamycin
   (3) Streptomycin
   (4) Tetracycline

Biochemistry

201. Radioactive labeling of RNA can best be achieved by using tritiated

(1) thymine
(2) adenine
(3) deoxyribose
(4) uracil

202. The reduction of purine ribonucleotides to purine deoxyribonucleotides may be said to

(1) utilize reducing equivalents whose ultimate source is NADPH
(2) involve the same reductase system for all four nucleosides
(3) take place at the level of the ribonucleoside diphosphate
(4) involve cleavage of the glycosidic bond in the sugar nucleotide

203. The nitrogens of the purine ring are derived from

(1) aspartic acid
(2) glycine
(3) glutamine
(4) serine

204. Feedback inhibition of the biosynthesis of purine nucleotides is effected by

(1) inosine nucleotides
(2) adenine nucleotides
(3) guanine nucleotides
(4) uric acid

205. Uric acid is a breakdown product of which of the following purine nucleotides?

(1) AMP
(2) GMP
(3) IMP
(4) CMP

206. Nucleoprotein complexes are constituents of

(1) viruses
(2) the translational complex
(3) ribosomes
(4) cell membranes

207. Which of the following statements about heterogenous RNA of the nucleus (hnRNA) are true?

(1) They have shorter lifetimes (a few minutes) than most cytoplasmic RNAs
(2) They have a long tail of poly A at the 3' end
(3) They are found in the extranucleolar portion of the nucleus
(4) The formation of the poly A tail is inhibited by 3'-deoxyadenosine

208. DNA ligase of E. coli may be said to

(1) catalyze the formation of a phosphodiester bond between the DNA chains of a double-helical DNA molecule
(2) catalyze the formation of a phosphodiester bond between two single-stranded DNA chains
(3) require NAD^+ as an energy source
(4) require ATP as an energy source

209. Which of the following statements about <u>DNA polymerase I</u> from *E. coli* are true?

(1) The enzyme can hydrolyze single-stranded DNA progressively from the 3′-hydroxyl terminus
(2) The enzyme has 5′-3′ nuclease activity in a double-helical region
(3) The enzyme requires a free 3′-OH in the DNA template
(4) The enzyme requires a free 5′-OH in the DNA template

210. *E. coli* grown on glucose-containing medium exhibits very low levels of a number of catabolic enzymes, e.g., β-galactosidase. Mechanisms involved in the inhibitory effect of glucose are described by which of the following statements?

(1) The *lac* operon is repressed by the binding of adenosine 3′,5′-monophosphate (cyclic AMP) to the promoter site of the *lac* operon
(2) Glucose reduces the intracellular concentration of cyclic AMP
(3) The *lac* operon promoter site recognizes a glucose-cyclic AMP complex
(4) A complex of cyclic AMP and a receptor protein binds to the promoter site of the *lac* operon

211. Operator genes have the ability to

(1) synthesize repressors
(2) affect structural genes
(3) synthesize aporepressors
(4) be affected by repression

212. Studies of the genetic code in prokaryotes have revealed that

(1) the nucleotide on the 3′-end of a triplet has the least specificity for an amino acid
(2) no signal exists to indicate the end of one codon and the beginning of another
(3) only three triplets are "nonsense" triplets
(4) mRNA molecules specify more than one polypeptide chain

213. Which of the following methods may be employed in "<u>nearest neighbor</u>" analysis of RNA?

(1) Synthesis of RNA with ^{14}C-labeled ribonucleotide triphosphates
(2) Alkaline hydrolysis of RNA products
(3) Hybridization to complementary DNA
(4) Synthesis of RNA with only one of the four ribonucleotide triphosphates labeled with ^{32}P at the alpha position

214. Which of the following are correct <u>Watson-Crick</u> base pairs in a characteristic DNA double helix?

(1) Adenine-thymine
(2) Uracil-adenine
(3) Cytosine-guanine
(4) Guanine-adenine

215. Which of the following statements about translation are true?

(1) Three nucleotide bases code for
one amino acid
(2) Specific nucleotide sequences
signal peptide chain termination
(3) The last nucleotide in a codon
has less specificity than the others
(4) More than one group of nucleotides may code for a single amino
acid

216. Which of the following statements about cell fusion are true?

(1) It occurs between ova and sperm
(2) It is part of conjugation in *E. coli*
(3) It allows for genetic recombination
(4) It is required for log phase growth

217. During the course of protein
synthesis, which of the following
events occur?

(1) Amino acids are attached
randomly to tRNAs
(2) The nascent peptide chain is
synthesized from its carboxyl
terminus
(3) The mRNA is pulled along the
ribosomes by contractile microtubules
(4) The growing peptide chain is
attached to a tRNA

DIRECTIONS: The groups of questions below consist of lettered choices followed by several numbered items. For each numbered item select the **one** lettered choice with which it is **most** closely associated. Each lettered choice may be used once, more than once, or not at all.

Questions 218-224

For each disorder that follows, select the mode of inheritance with which it is most likely to be associated.

(A) Autosomal dominant inheritance
(B) Autosomal recessive inheritance
(C) X-linked dominant inheritance
(D) X-linked recessive inheritance
(E) None of the above

A 218. Acute intermittent porphyria

C 219. Vitamin D-resistant rickets with hypophosphatemia

D 220. Lesch-Nyhan syndrome ✓

B 221. Phenylketonuria

A 222. Familial hypercholesterolemia

D 223. Red-green color blindness ✓

B 224. Wilson's disease

Questions 225-228

For each of the RNA species that follow, choose the description with which it is most likely to be associated.

(A) Part of the ribosomal subunit with peptidyl transferase acitvity
(B) Part of the ribosomal subunit that binds mRNA
(C) Part of the chromosome
(D) Carries the anticodon and serves as adaptor to translate the genetic code into a sequence of amino acids
(E) Complementary to the base sequence of an operon and codes, therefore, for the polypeptide chains that correspond to the genes of the operon

E 225. mRNA

D 226. tRNA

A 227. 23-S rRNA

B 228. 16-S rRNA

Nucleic Acids and Molecular Genetics

Answers

111. The answer is D. *(Stryer, pp 557-558, 595.)* The bases found in DNA are adenine, guanine, cytosine, and thymine, whereas RNA contains uracil in place of thymine. Thus an animal fed tritiated thymine would incorporate label into DNA and not RNA. Tritiated adenine, guanine, and cytosine would be incorporated into both polymers.

112. The answer is D. *(Stryer, pp 564-595.)* Adenine, guanine, cytosine, and uracil are found in all RNA species. DNA contains thymine in place of uracil. In double-stranded DNA, thymine pairs with adenine in the complementary strand and during transcription mRNA is formed on the DNA template using uracil to pair with the adenine in the readout strand of DNA.

113. The answer is C. *(Stryer, p 652.)* Up to 10 percent of the nucleotides of tRNA have unusual methylated bases. Mitochondrial and nucleolar DNAs contain conventional bases. Nucleolar DNA codes for rRNA.

114. The answer is A. *(White, ed 6. pp 779, 1130.)* The Lesch-Nyhan syndrome is caused by the absence of the enzyme hypoxanthine-guanine phosphoribosyl-transferase (HGPRT). Symptoms of the Lesch-Nyhan syndrome include hyperuricemia and gout. Other features of the syndrome such as mental deficiency and self-mutilation are less well explained by HGPRT deficiency.

115. The answer is E. *(Mahler, ed 2. pp 806-807, 833-834. White, ed 6. p 694.)* S-Adenosylmethionine donates methyl groups in the biosynthesis of creatine phosphate, epinephrine, melatonin, and phosphatidylcholine. Thymine, however, receives a methyl group from N^5, N^{10}-methylenetetrahydrofolate.

116. The answer is B. *(White, ed 6. pp 694-695.)* S-Adenosylmethionine donates the methyl group of methionine in many biochemical reactions, leaving S-adenosylhomocysteine. These reactions include methylation of guanidoacetic acid to yield creatine, and of phosphatidylethanolamine to yield phosphatidylcholine.

117. The answer is D. *(White, ed 6. pp 777-778.)* The final step of purine catabolism in primates is the oxidation of xanthine to uric acid. Overproduction or undersecretion of uric acid causes the urate crystal formation of gout. Urea, arising from allantoic acid, and allantoic acid itself are the excreted products of purine catabolism in fish, but in mammals urea arises from the catabolism of amino acids. Xanthine and orotic acid, which are not normally excreted, are intermediates in purine and pyrimidine metabolism respectively.

118. The answer is A. *(White, ed 6. p 757.)* Purine synthesis is a metabolic pathway that is common to bacteria and humans. Nitrogen fixation, photophosphorylation, mucopeptide synthesis, and fermentation occur in plants or bacteria but not in humans. Thus, photophosphorylation, which is equivalent to oxidative phosphorylation in animals, involves ATP formation during photosynthesis.

119. The answer is A. *(Stryer, p 532.)* Carbons 4 and 5 and nitrogen 7 of purines are derived from glycine. The glycine molecule is incorporated intact into the purine ring. Carbons 2 and 8 arise from activated derivatives of tetrahydrofolate; carbon 6 arises from carbon dioxide.

120. The answer is C. *(Stryer, pp 532-535, 538-539.)* An early step in pyrimidine synthesis is the reaction catalyzed by aspartate transcarbamoylase: carbamoylphosphate + aspartate → carbamoyl aspartate. Carbamoyl phosphate is not an intermediate in purine biosynthesis. Features, A, B, and D listed in the question are common to the biosynthetic pathway of purine biosynthesis; feature E is true of pyrimidine as well as of purine biosynthesis.

121. The answer is B. *(White, ed 6. pp 697-699.)* Mitochondrial carbamoyl phosphate synthetase is involved in urea biosynthesis (via arginine hydrolysis); cytoplasmic carbamoyl phosphate synthetase is involved in pyrimidine biosynthesis. Both enzymes occur in the liver of mammals. The mitochondrial enzyme is present in relatively high activity and is activated by acetylglutamate, and the cytoplasmic enzyme is present in low activity and is inhibited by uridine triphosphate.

122. The answer is A. *(Stryer, pp 538, 544.)* The pyrimidine ring is formed from carbamyl phosphate and aspartic acid. The former arises from the reaction of glutamine and HCO_3^- with carbon A of the diagram derived from the HCO_3^- and nitrogen B from the amide N of glutamine. Nitrogen D and carbons A and E are derived from aspartic acid, with carbon A being the side-chain carboxyl carbon. Folic acid derivatives are **not** required in the synthesis of the pyrimidine ring, but 5,10-methylenetetrahydrofolic acid **is** required for the synthesis of thymine from uracil.

123. The answer is D. *(White, ed 6 pp 226-227, 764-765.)* In bacteria, cytidine triphosphate leads to feedback inhibition of aspartate transcarbamoylase. Aspartate transcarbamoylase has been shown to consist of 12 subunits, six each of catalytic and regulatory subunits. In mammals, uridine triphosphate inhibits pyrimidine biosynthesis at the level of the cytoplasmic carbamoyl phosphate synthetase rather than at the level of aspartate transcarbamoylase.

124. The answer is D. *(White, ed 6. pp 188-189.)* Bovine pancreatic ribonuclease cleaves RNA at a point distal to the $3'$ phosphate of a pyrimidine nucleotide. The products, therefore, are a pyrimidine nucleoside $3'$-phosphate and oligonucleotides with a terminal pyrimidine $3'$-phosphate. RNA hydrolysis by snake venom phosphodiesterase, however, yields mainly nucleoside $5'$-phosphates.

125. The answer is C. *(Mahler, ed 2. pp 217-220.)* The sedimentation coefficient (s^0) of a DNA molecule is given by the formula $s^0 = MD(1-\bar{v}\rho)/RT$, where M is the molecular weight, $\bar{v}$ is partial specific volume, and ρ is the density (mass/volume) of the solvent. D, the diffusion coefficient, is determined in part by the molecular shape. R is the gas constant and T is temperature ($^\circ K$). Optical density has nothing to do with the sedimentation coefficient; it measures the light absorbance of a solution.

126. The answer is B. *(Mahler, ed 2. pp 235-241.)* The melting temperature of DNA is defined as the temperature at which half the base pairs of DNA have been disrupted—that is, half the DNA has undergone the structural change referred to in protein chemistry as denaturation. One way of assaying the denaturation is to measure the optical density of the DNA (usually at 260 nm). Denatured DNA has an optical density approximately 40 percent greater than native DNA.

127. The answer is D. *(Mahler, ed 2. pp 204, 238-239.)* The melting temperature (T_m) of duplex DNA is the temperature at which half the base pairs are denatured. Because cytosine-guanine (C-G) base pairs have three rather than two hydrogen bonds, a high content of C + G increases T_m by virtue of the positive straight-line relationship. According to Chargaff's rules, in duplex DNA the content of A + G = C + T.

128. The answer is B. *(Stryer, pp 563-566.)* The Watson-Crick model of DNA structure predicts a double-stranded helix with hydrogen bonds between A-T and G-C bases on the inside of the helix. The chains run in an antiparallel, or opposite, direction. Covalent phosphodiester bonds exist only between sugar moieties within each chain and not between their attached bases.

129. The answer is E. *(Mahler, ed 2. pp 256-258. White, ed 6. p 187.)* Histones are a family of basic proteins (rich in lysine and arginine) that are bound non-covalently but stoichiometrically to DNA. They are of relatively low molecular weight (10,000-20,000 daltons) and separable into five classes, largely on the basis of their lysine and arginine content. Histone IV from calf thymus and pea buds discloses a remarkable degree of homology: these diverse organisms exhibit a difference in only 2 out of 102 amino acids that constitute their protein.

130. The answer is A. *(White, ed 6. pp 176-178.)* DNA replication entails pairing of thymine with adenine and guanine with cytosine. The chains of the double helix are thus bonded in part by a hydrogen linkage between amino and keto groups. The strands themselves are synthesized in an antiparallel direction, i.e., the $5' \rightarrow 3'$ sequence of phosphodiester bridges mentioned in the question specifies its complement in a $3' \rightarrow 5'$ direction.

131. The answer is E. *(White, ed 6. p 821.)* AUG is the codon for both methionine and N-formylmethionine. In *E. coli*, AUG is the chain-initiating codon and N-formylmethionine is the first amino acid incorporated into the nascent polypeptide. In mammals, AUG is also thought to be the codon for chain-initiation, but methionine (rather than N-formylmethionine) is the N-terminal amino acid.

132. The answer is C. *(Lehninger, ed 2. pp 860-862.)* Because it is thought that the genes are constant in all cells of a given organism (with the exception of antibody-producing cells), the ratios of nucleotide bases, (A+T)/(G+C), also must be constant in these cells. These ratios may vary from species to species but should not vary with age or environmental factors. The differentiation of one tissue from another, in any single animal, depends upon the variable expression of a genetic code common to all the tissues of that animal.

133. The answer is A. *(White, ed 6. pp 787-788.)* Semiconservative double-stranded DNA replication demands that from each half (strand) of the label-led parental DNA, a complement will be synthesized that maintains the parental structure intact. Therefore, the first round of replication in a cold (unlabeled) solution will yield two molecules that are half-labeled. The second round will yield two half-labeled and two unlabeled molecules of double-stranded DNA.

134. The answer is E. *(White, ed 6. pp 831-832.)* Insertion of one extra nu-cleotide causes a frame-shift mutation and mistranslation of all the mRNA transcribed from beyond that point in the DNA. All the other mutations cited in the question usually cause an error in the identity of only one amino acid (A or B), or removal of one amino acid from the sequence D, or no error at all in the amino acid sequence C. There is a chance that either A or B will give a "nonsense", or chain-terminator, mutation, and this is about as likely to be lethal as is a frame-shift.

135. The answer is D. *(Mahler, ed 2. pp 940-949. White, ed 6. pp 839-840.)* The anticodon CGU is complementary to the codon ACG. Note that both trip-lets are written with the 5'-end at the left and the 3'-end at the right: Watson-Crick base pairing is antiparallel. The anticodon site in tRNA frequently contains unusual nucleotides, such as inosinic acid or pseudouridylic acid.

136. The answer is B. *(White, ed 6. pp 798-799.)* Dimerization between thymines on the same polynucleotide strand is caused by ultraviolet radiation (260 nm) and stops replication at that point. The dimers can be excised and repaired by enzymes that include ligase, or can be dissociated by further ex-posure to longer (330-450 nm) or shorter (230 nm) wavelengths in the process of photoreactivation.

137. The answer is E. *(White, ed 6. pp 829, 831-832, 975.)* In HbS, a valine residue replaces a glutamic acid on the β chain as the result of a point mutation in one nucleotide base. This single nucleotide alteration at the second position of the triplet consists in a change of thymine to adenine.

138. The answer is C. *(White, ed 6. pp 836-837.)* *N*-Formylmethionine is thought to be always the first amino acid at the N-terminus of polypeptides synthesized in *E. coli*. Two discrete tRNAs for methionine exist in *E.coli* to specify whether or not *N*-formylation is to occur.

139. The answer is A. *(Lehninger, ed 2. pp 963, 970. Watson, ed 3. pp 357-359.)* According to the wobble hypothesis of Crick, the base pairing at the third position of the codon is less stringent than at the other two positions. Thus, certain bases at the 5'-position of the anticodon are able to pair with more than one base at the 3'-position of the codon. This allows one tRNA species to read more than one codon.

140. The answer is C. *(White, ed 6. p 823.)* The chain-terminating (nonsense) codons are UAA, UAG, and UGA. Termination of a polypeptide chain occurs when one of these codons is read in phase, and the tRNA bound to the C-terminal residue of the polypeptide is liberated.

141. The answer is B. *(Stryer, pp 603-606.)* RNA polymerase, which requires a DNA template in order to synthesize a ribonucleic acid polymer, usually transcribes one strand of double-helical DNA. The direction of growth of the RNA chain is from the 5'- to the 3'- end, and the product is never a circular molecule. Unlike DNA polymerase, RNA polymerase does not require a primer.

142. The answer is A. *(Mahler, ed 2. pp 905-907.)* Sigma factor is the subunit of RNA polymerase that confers specificity of initiation on the core enzyme. In the presence of sigma factor, RNA polymerase will choose the correct strand of duplex DNA for transcription and will initiate transcription at the appropriate promoter region.

143. The answer is D. *(Lehninger, ed 2. pp 931-932.)* The first step in the incorporation of amino acids into polypeptide chains is known as the activation step, which is catalyzed by aminoacyl-tRNA synthetases. The reaction involves the esterification of amino acids to their specific tRNA at the expense of ATP. Pyrophosphate formed in the reaction is hydrolyzed rapidly by pyrophosphatase, which ensures the rapid completion of the reaction. Other steps include initiation, involving the binding of mRNA and the first aminoacyl tRNA to the 30-S subunit of the ribosome; elongation, in which the polypeptide chain is lengthened by addition of another aminoacyl tRNA; and termination, which releases the polypeptide from the ribosome and is defined by special stop codons on the mRNA.

144. The answer is A. *(White, ed 6. pp 836-837.)* Two molecules of guanosine triphosphate (GTP) are hydrolyzed in the process of forming one peptide bond by the ribosome. Ribosomal peptidyl transferase is located in the 50-S subunit. The three distinct soluble peptide elongation factors are the transfer factors Tu and Ts, and the protein G. The protein G possesses ribosome-dependent GTPase activity.

145. The answer is C. *(Stryer, p 672.)* Puromycin is an analog of the 3'-end of an aminoacyl-tRNA. It binds to the A site on the 50-S subunit of the ribosome, inhibiting the entry of aminoacyl-tRNA, and its α-amino group forms a peptide link with the carboxyl group of the nascent polypeptide chain. The peptide, with puromycin covalently bound to its carboxyl end, is released from the ribosome. Puromycin has been particularly useful in the development of the concept of A and P sites on the ribosome.

146. The answer is D. *(Stryer, p 671.)* Tetracycline blocks the binding of aminoacyl-tRNA with initiator sites on 30-S subunits of ribosomes, causing an irreversible inhibition of protein synthesis. The transcription of DNA, the binding of mRNA to ribosomes, and the formation of aminoacyl-tRNAs are not affected. Release of peptides from mRNA-tRNA complexes does not occur inasmuch as peptide bonds are not formed.

147. The answer is E. *(Stryer, pp 670-671.)* Chloramphenicol interacts with the 50-S subunit of prokaryotic ribosomes to inhibit the process of polypeptide chain elongation. Streptomycin, on the other hand, binds to the 30-S subunit of ribosomes resulting in decreased rates of protein synthesis, but additionally it induces misreading of mRNA codons.

148. The answer is C. *(Lehninger, ed 2. p 879.)* 5-Bromouracil is inserted into DNA during replication because it has a close structural resemblance to thymine. However, tautomerization of the keto to the enol form occurs more easily than with thymine, with the enol form pairing with G, instead of A. Thus, during replication, a C is inserted into the complementary strand in place of a T.

149. The answer is A. *(White, ed 6. p 815.)* During translation, messenger RNA (mRNA) codons dictate the amino acids to be synthesized into proteins. This is in contrast to transcription, which **yields** mRNA complementary to a sequence of DNA, along with ribosomal RNA (rRNA) and transfer RNA(tRNA).

150. The answer is B. *(Lehninger, ed 2. pp 939-941.)* The two subunits of ribosomes are composed of proteins and rRNA. Ribosomes are found in the cytoplasm, in mitochondria, and bound to the endoplasmic reticulum. Transcription refers to the synthesis of RNA complementary to a DNA template, and has nothing immediately to do with ribosomes.

151. The answer is D. *(White, ed 6. pp 181-182.)* Guanine-cytosine (G-C) base pair bonds are stronger than adenine-thymine (A-T) bonds. DNA with a high G-C content is more stable and melts at a higher temperature than DNA that is rich in A-T. This stability difference is a consequence of the existence of only two hydrogen bonds between A and T, compared to three between G and C.

152. The answer is D. *(White, ed 6. pp 821-825.)* The initiation but not the elongation of polypeptide chain synthesis (in *E. coli*) involves N-terminal *N*-formylmethionine incorporation as specified by codons **AUG** and **GUG**. Peptide chain elongation upon the ribosomal mRNA employs peptidyl transferase after codon-specific binding of aminoacyl tRNA, using energy from GTP and the factors G, Tu, and Ts.

153. The answer is C. *(Davis, ed 2. pp 107-108.)* The cell walls of gram-positive bacteria exhibit a 200 to 800 Å-thick, electron-dense outer layer consisting of *N*-acetylmuramic acid, *N*-acetylglucosamine, and **D**- and **L**-amino acid peptides. The enzyme lysozyme partially hydrolyzes cell walls.

154. The answer is C. *(Jawetz, ed 13. pp 44, 51.)* Transformation is characterized by the bacterial uptake of soluble DNA released by a donor cell. Transformation takes place only in bacteria that can utilize the high molecular weight DNA of the medium. Although originally discovered in the pneumococcus, transformation also occurs in *Hemophilus, Bacillus,* and *Neisseria* species, as well as in *E. coli*.

155. The answer is C. *(Davis, ed 2. pp 1113-1118.)* Temperate phages can transfer genetic material from one cell to another in a process known as transduction. In generalized transduction, any of the bacterial genes may be transferred together with part of the phage genome, whereas in specialized transduction only a small part of the bacterial genome, restricted to genes adjacent to the point of insertion of the phage in the bacterial genome (the prophage site), are transferred. The classic phage shown to cause specialized transduction in *E. coli* is phage λ, which transfers the gal (galactose utilization) or bio (biotin synthesis) regions of the *E. coli* genome.

156. The answer is A. *(Jawetz, ed 13. pp 44-45.)* Transduction is the process in which a piece of donor chromosome is carried to the recipient by a temperate bacteriophage produced by the donor cell. There are two types of transduction. In the generalized type, the phage has an approximately equal chance of transducing any segment of the donor's chromosome. In restricted transduction, the phages carry only segments that are immediately adjacent to the prophage.

157. The answer is D. *(Jawetz, ed 13. pp 43-52.)* Chromosomal material can be transferred between bacterial cells by transformation, transduction, and conjugation. Large segments of chromosomes, however, are passed only through conjugation.

158. The answer is E. *(Stryer, pp 572-576, 602-605.)* Nucleoside triphosphates are the substrates for both RNA and DNA polymerases, and both enzymes add bases at the free $3'$-end of the growing polynucleotide chain. DNA polymerase synthesizes DNA that is complementary to DNA. The enzyme that synthesizes DNA that is complementary to RNA is called reverse transcriptase.

159. The answer is E. *(Davis, ed 2. pp 1174-1177.)* Interferon is a cell-specific protein produced by cells infected by a virus. Interferon elicits production of an antiviral protein both in cells that produce it and in those that are exposed to it. While all animal cells are capable of yielding interferon, cells of the reticuloendothelial system are the major source in an infected animal. Interferon production is dependent on transcription and translation.

160. The answer is A. *(Jawetz, ed 13. pp 39-41.)* Spontaneous point mutations are most likely to arise from tautomeric shifts of hydrogen atoms in a purine or pyrimidine base. This shift causes an altered base-pairing during DNA replication. Several mutagenic agents such as 5-bromouracil and 2-aminopurine increase tautomerization of bases in DNA.

161. The answer is A. *(Davis, ed 2. pp 256-258.)* Frame-shift mutations caused by the insertion or deletion of a single base pair are induced by acridine derivatives. 5-Bromouracil produces transitional mutations, because substitution of bromine for the 5-methyl group of thymine increases the likelihood that the enol tautomeric form will undergo base-pairing with guanine instead of adenine. Azathioprine is converted into 6-mercaptopurine, a purine analog. Ethyl ethanesulfonate produces transitions through alkylation of guanine.

162. The answer is B. *(McGilvery, p 387.)* Humans cannot synthesize phenylalanine, therefore it is always an essential amino acid. Because phenylketonuric patients cannot hydroxylate phenylalanine to form tyrosine, in these individuals tyrosine also is an essential amino acid.

163. The answer is E. *(Stryer, pp 681-683.)* Transcription of the lactose operon is initiated by binding of RNA polymerase (including sigma factor) to the promotor region, "p." The region designated "i" governs repressor synthesis, and repressor protein from "i" is the negative control exerted upon the operon at the site designated "o." The regions "z," "y," and "a" specify β-galactosidase, galactoside permease, and acetylase enzymes, respectively.

164. The answer is B. *(White, ed 6. pp 813-814.)* The smallest unit of DNA capable of coding for the synthesis of a polypeptide is the cistron. A gene sequence under the coordinated control of a single operator is called an operon. The promoter is the site of binding of RNA polymerase and initiation of transcription, and a replicon is a unit of DNA that contains a signal for initiation of DNA synthesis.

165. The answer is E. *(Jawetz, ed 13. pp 46-53.)* Chromosome transfer can occur by conjugation between bacterial cells containing the F-factor (F⁺) and those without an F-factor (F⁻). If the F-factor is integrated with the bacterial genome in the donor cell, the cells are called Hfr (high frequency of recombination), and in the presence of an excess of F⁻ cells all Hfr cells will initiate replicative transfer with an F⁻ cell. The initiation point of transfer of the bacterial chromosome is determined by the point of integration of the F-factor with the bacterial chromosome. In *E. coli* K12 there are several preferred sites of integration of F-factor and bacterial genome, apparently specified by regions of base-pair homology.

166. The answer is A. *(Hayes, ed 2. p 48.)* Since double crossovers between two genetic loci result in no apparent recombination of the two loci, the recombination frequency is always less than the sum of the intervening frequencies. In the absence of double crossovers, the recombination and crossover frequencies would exhibit a direct linear relationship.

167. The answer is A. *(Hayes, ed 2. p 55.)* Genetic transfer can occur in bacteria by three different mechanisms. In one of these, called transformation, high molecular weight DNA extracted from donor cells can induce hereditary changes in recipient cells. The size of the fragments involved in the transformation is about one percent or less of the bacterial (donor) chromosome.

168. The answer is D. *(Miller, p 64.)* Of the bacterial characteristics listed in the question, only the F pili have been associated with sexual activity. The transfer of the bacterial chromosome, the formation of recombinants, and the other properties mentioned do not necessarily occur during bacterial conjugation.

169. The answer is D. *(Lehninger, ed 2. p 870.)* Mitochondria of eukaryotic cells contain four to six molecules of circular DNA, as shown in the electron micrograph accompanying the question. The nuclear DNA of eukaryotes and the *E. coli* DNA genome are much larger. Lysosomes do not contain DNA.

170. The answer is B. *(Miller, p 82.)* The cells containing the F-factor (male cells) and which are sensitive to streptomycin (str^s) will be killed on streptomycin plates (i.e., either B or C in the question). If lactose-minimal streptomycin plates are used, the only female cells that will grow will be those that received the F-factor and the lac$^+$ genes. Thus a lac$^+$ str^r strain can be selected.

171. The answer is E. *(Lewin, pp 51, 126, 159-162, 500-501, 565.)* Puromycin is a structural analog of the aminoacyl end of the tRNA. It irreversibly reacts with the peptidyl-tRNA, thereby terminating protein synthesis. Streptomycin, like tetracycline and chloramphenicol, inhibits ribosomal activity. Mitomycin covalently cross-links DNA, preventing cell replication. Rifampicin is an inhibitor of DNA-dependent RNA polymerase.

172. The answer is B. *(Lewin, p 15.)* In a fully overlapping genetic code, a single base mutation would alter three amino acids: in a non-overlapping code only one amino acid is altered. With either coding system, more than one triplet might specify a given amino acid. Ribosomal binding and termination triplets have no bearing upon deciphering of the coding system. The fact that poly (U-G) directs the synthesis of poly(cys-val) is ambiguous in the deciphering of genetic coding.

173. The answer is C. *(Lewin, p 5.)* The major implication of the *cis/trans* test is that when two strains have mutant sites lying in the same gene, they show complementation to give a wild phenotype in only the *cis* arrangement. Two strains having mutant sites in different genes complement in either the *cis* or *trans* arrangement.

174. The answer is E. *(Lewin, p 505.)* Xeroderma pigmentosum appears to be due to the inability of an excision-repair system to remove thymine dimers, which are formed upon exposure of DNA to ultraviolet radiation. Mutagenesis by this mechanism is presumably the basis for the multiple neoplasms that occur in patients who have this disease.

175. The answer is A. *(Miller, p 79.)* If an $F'lac^+$ factor is carried in a lac strain, cells which have lost the F' factor will not grow, as they must subsist in a lactose medium. The use of a $recA^-$ strain will prevent any incorporation of the $F'lac^+$ into the bacterial chromosome, as it is deficient in recombination.

176. The answer is E. *(Lewin, p 280. Stryer, pp 681-683.)* The term operon refers to the system of operator gene together with the structural genes that are transcribed in concert when the operator gene is active. "Operon" includes the promoter gene (the site of attachment of RNA polymerase) but not the regulatory gene (site of coding for repressor). Codons comprise a linear array of three nucleotide bases and are the simplest unit of any genetic message. Cistrons are functional genetic units defined by complementation analysis and a muton is the smallest unit defined by mutational analysis. The genome is the entire genetic message of a species.

177. The answer is E. *(Lewin, pp 44, 76.)* The synthesis of an aminoacyl-tRNA requires two high-energy phosphate bonds, one for the formation of the ester linkage and the other for driving the reaction forward. In addition, two molecules of GTP are hydrolyzed, one in the binding of the aminoacyl-tRNA to the A site on the ribosome, and the other for translocation of the peptidyl-tRNA from the A site to the P site. Thus four high-energy bonds are expended in the process.

178. The answer is D. *(Lewin, p 414.)* 5-Bromouracil is incorporated into DNA, in place of thymidine, to yield a denser DNA. The newly synthesized DNA fragments can then be quantitated by centrifugation through density gradients of cesium chloride. 5-Bromouracil is neither more reactive nor more sensitive to cleavage than thymidine, nor does it cause frame-shift mutations, as do the acridine dyes.

179. The answer is E. *(Miller, pp 73-74.)* The transfer of a prophage to a phage-sensitive recipient is called zygotic induction. If Hfr (high frequency of recombination) strains containing repressible prophage are mated with sensitive cells that are not lysogenic (i.e., lack immunity to lysis by phage), the transferred phage will be replicated in the sensitive cell, leading to phage burst (cell lysis). Donor chromosomal genes that are transferred after transfer of the prophage region of the genome will appear with low frequency in the recipient cells compared to those genes transferred before the prophage.

180. The answer is B. *(Miller, p 115.)* Acridines induce frame-shift mutations, in which a nucleotide is either added to or removed from the DNA and the mRNA transcribed from it will correspondingly be lengthened or shortened. Since mRNA is translated as triplets of nucleotides without punctuation, the sequence of amino acids laid down after the point of the frame shift will bear no resemblance to the unmutated sequence. The mechanism of this frame shift appears to be intercalation of the dye between adjacent base pairs in the DNA double helix. If the dye remains intercalated, reading is advanced, but if the dye and adjacent base are deleted, reading is retarded.

181. The answer is C. *(White, ed 6. pp 884-889. Stryer, pp 712-713.)* Infection of an *E. coli* cell by T4 phage very rapidly leads to loss of host DNA and other macromolecular synthesis. Viral genome-specified deoxyribonucleases are formed immediately upon infection and these destroy host DNA. The phage DNA is not destroyed because it contains hydroxymethyl cytosine in place of cytosine, and is also glucosylated. The phage deoxyribonuclease recognizes clusters of cytosines.

182. The answer is C. *(Lewin, p 345.)* Lysogeny is maintained by the synthesis of viral regulator proteins that prevent the expression of phage genes needed for lytic development. The same regulators repress any other phage DNA of the same type and, therefore, confer immunity.

183. The answer is B. *(Lewin, p 310.)* The *lac* operon is shut off by a repressor protein that is inactivated by an inducer—i.e., negative control of induction. In the absence of an inducer, transcription of the galactosidase structural gene into mRNA cannot proceed. The combination of repressor protein and inducer constitutes a repressor-induced complex.

184. The answer is E. *(Hood, pp 6-8.)* In the positive-control-of-induction model, it is postulated that the regulator gene specifies an inducer protein that requires activation by an inducer molecule before it will interact with the operon to switch it on. An example of this type of control is found in the arabinose *ara* operon of *E. coli.*

185. The answer is E. *(Lewin, p 280.)* The repressor protein binds to a specific operator site on the operon. This prevents the RNA polymerase from transcribing the subsequent structural genes into mRNA. Repressor protein does not act distally in this mechanism, either to prevent translation or actual protein synthesis.

186. The answer is D. *(Hayes, ed 2. p 174.)* The distribution of spontaneous mutations observed in the T4 phage genome is far from random. Some sites in the r_{II} region are much more likely to be a site of change than others; these areas of the genome Benzer called hot spots.

187. The answer is C. *(Hayes, ed 2: p 56.)* Transduction involves the use of bacterial viruses to transfer small regions of a bacterial chromosome from a donor to a recipient. Although of little use for detecting linkage between genes that are not close together, transduction can be a valuable method for the analysis of genetic fine structure.

188. The answer is E (all). *(Lehninger, ed 2. pp 931-932, 941-942.)* ATP is required for activation of amino acids for the formation of aminoacyl-tRNAs that interact with ribosomes carrying mRNA. Peptidyl transferase catalyzes the formation of peptide bonds between the free amino group of the aminoacyl-tRNA on the A site of the ribosome and the esterified carboxyl group of the peptidyl-tRNA on the P site, the liberated tRNA remaining on the P site.

189. The answer is B (1, 3). *(White, ed 6. pp 173-179.)* The sugar moiety of DNA is 2-deoxyribose where H has replaced the OH in the $2'$ position making it impossible to form a phosphate bond at this position. There are no phosphate → phosphate bonds in DNA (as exist in ADP and ATP). The main chain of the two antiparallel strands of a DNA molecule consists of deoxyribose residues linked by covalent phosphodiester linkages from $3'$- to $5'$-positions.

190. The answer is B (1, 3). *(White, ed 6. pp 833, 836.)* Rifampicin blocks the initiation of DNA-dependent RNA synthesis, but not the elongation of chains that already have been initiated. Actinomycin D blocks elongation; by complexing with guanine residues in DNA, it inhibits all RNA synthesis and thereby protein synthesis. Chloramphenicol and tetracycline inhibit protein, but not nucleic acid, synthesis.

191. The answer is D (4). *(Jawetz, ed 13. pp 49-50.)* DNA transfer by Hfr donors occurs when a suspension of Hfr cells is mixed with an excess of F^- cells. At random times, transfer will be interrupted by the spontaneous breakage of the DNA molecule. The Hfr state is reversible; detachment of the F factor takes place approximately once per 10^5 cells at each generation.

192. The answer is B (1, 3). *(Jawetz, ed 13. pp 46-53.)* Bacterial conjugation requires physical contact between donor and recipient cells joined by an F pilus. Small or large segments may be transferred, depending on the state of the F^+ factor. F^- cells can act only as recipients, while F^+ cells may behave as donors or recipients.

193. The answer is E (all). *(Jawetz, ed 13. pp 49, 53.)* Bacteria acquire drug resistance by spontaneous mutation, recombination, or transduction. Resistance to penicillin in gram-positive staphylococci is plasmid-determined. However, conjugal transfer does not occur in gram-positive bacteria and the plasmids are transferred by transducing phages.

194. The answer is B (1, 3). *(Jawetz, ed 13. pp 101-102.)* Early proteins synthesized in response to phages include enzymes necessary for the synthesis of new phage DNA: DNA polymerase, thymidylate synthetase, and kinases for the formation of nucleoside triphosphates. The T2, T4, and T6 phages that incorporate hydroxymethylcytosine into their DNA cause the host cell to synthesize an additional series of enzymes needed for the synthesis of hydroxymethylcytosine.

195. The answer is A (1, 2, 3). *(Jawetz, ed 13. pp 50-52.)* High-frequency recombination (Hfr) strains of bacteria integrate the F agent into their chromosome. If conjugation of an Hfr cell with an F^- cell occurs, replicative transfer commences at the site of integration of the F factor with the chromosomal DNA, but it is the F-attached chromosome rather than the F agent that is transferred. All cells of the Hfr strains begin chromosome transfer at the same locus and produce an orderly sequence of gene entry and a rate of recombination that is higher for genes closer to the transfer locus than those more distant.

196. The answer is C (2, 4). *(Davis, ed 2. pp 120-121, 151.)* Bacitracin is a bacteriolytic cyclic peptide that blocks cell wall synthesis by inhibiting phosphorylation of undecaprenol-PP formed during the transfer of subunits from carrier to cell wall. Penicillin blocks cross-linking of the peptidoglycan (transpeptidation reaction). Chloramphenicol inhibits protein synthesis, while sulfonamides are competitive analogs of *p*-aminobenzoate.

197. The answer is E (all). *(Jawetz, ed 13. pp 46-47, 49.)* Bacteria contain extrachromosomal genetic elements known as plasmids that can mediate chromosome transfer via conjugation. The most widely studied plasmid is F, the sex factor of *E. coli* K12. Bacteria containing the sex factor may act as females or males and are able to conjugate with males or females.

198. The answer is C (2, 4). *(Jawetz, ed 13. pp 124-125.)* Chloramphenicol, a strong inhibitor of protein synthesis, is effective against many gram-negative rods, cocci, and rickettsiae. The antibiotic blocks amino acid attachment to nascent peptide chains by interfering with the action of peptidyl transferase.

199. The answer is A (1, 2, 3). *(White, ed 6. pp 186-192.)* Viruses can reproduce only inside living cells and are incapable of independent DNA replication. Prokaryotes (bacteria), as well as the nuclei and mitochondria of eukaryotic cells, are able to sustain DNA replication.

200. The answer is E (all). *(Mahler, ed 2. pp 953-957.)* All the compounds listed in the question block peptide synthesis by binding to ribosomes. The modes of action vary in detail. Tetracycline, for example, inhibits binding of aminoacyl tRNA to mRNA by interference at the level of the 30-S subunit of the ribosome. Streptomycin, by contrast, binds to the 50-S subunit of the ribosome, decreasing the rate of protein synthesis and also interfering with the reading of mRNA codons.

201. The answer is C (2, 4). *(White, ed 6. pp 167-170, 189.)* The pyrimidine thymine and the sugar deoxyribose are incorporated into DNA, not RNA, although thymine can exist as a ribonucleoside in tRNA. Adenine is incorporated into both RNA and DNA and uracil only into RNA.

202. The answer is A (1, 2, 3). *(Stryer, p 543.)* The biosynthesis of deoxyribonucleotides occurs by reduction of the $2'$ hydroxyl group of the pentose moiety in all four nucleoside diphosphates. The immediate hydrogen donor is reduced thioredoxin, which is regenerated by thioredoxin reductase and NADPH-coupled flavoprotein. Thus NADPH is the ultimate source of reducing equivalents.

203. The answer is A (1, 2, 3). *(Stryer, p 532.)* The ring nitrogens of purines are derived from glutamine (N-3, N-9), glycine (N-7), and aspartate (N-1). Carbon atoms 4 and 5 also are furnished by glycine to provide the carbon double-bond, backbone of the purine ring.

204. The answer is A (1, 2, 3). *(Stryer, p 537.)* Feedback inhibition of purine nucleotide biosynthesis occurs at several sites in the pathway. AMP, GMP, and IMP inhibit both the formation of 5-phosphoribosylpyrophosphate (PRPP) from ribose 5-phosphate and the conversion of PRPP to phosphoribosylamine. AMP inhibits the conversion of IMP to adenylosuccinate (AMP precursor) and GMP inhibits the conversion of IMP to xanthylate (GMP precursor). Uric acid is not a feedback inhibitor of purine nucleotide biosynthesis.

205. The answer is A (1, 2, 3). *(Harper, ed 16. p 392.)* Xanthine oxidase catalyzes the conversion of hypoxanthine or xanthine to uric acid. Purine nucleotides, like AMP, GMP, and IMP, are catabolized to uric acid via hypoxanthine and xanthine.

206. The answer is A (1, 2, 3). *(Lehninger, ed 2. pp 303, 331.)* Cell membranes are composed of lipid and protein and do not include DNA. The translational complex refers to the combination of mRNA, tRNA, and ribosomes involved in protein synthesis.

207. The answer is E (all). *(Hood, pp 184-185. Stryer, pp 705-706.)* Very large RNA molecules (hnRNA) are found in the extranuclear portion of the nucleus of eukaryotic cells and have lifetimes much shorter than many cytoplasmic RNAs. Most hnRNAs contain a poly A tail of about 150-200 residues at the 3' end. The inhibitor 3'-deoxyadenosine, which blocks the synthesis of the 3'-terminal poly A tail without inhibiting general hnRNA synthesis, prevents the appearance of new mRNA in the cytoplasm.

208. The answer is B (1, 3). *(Stryer, pp 577-578.)* DNA ligase catalyzes the formation of a phosphodiester bond between two DNA chains that must be part of a double-stranded DNA molecule; the enzyme cannot link single-stranded DNA molecules. The energy source for the bond formation in *E. coli* is NAD^+; in some animal cells and bacteriophage, ATP serves this purpose. DNA ligase functions in DNA synthesis and repair, and in recombination.

209. The answer is A (1, 2, 3). *(Lehninger, ed 2. pp 895-897.)* The primary function of DNA polymerase I from *E. coli* is to catalyze the step-by-step addition of a deoxyribonucleotide from a deoxyribonucleoside triphosphate to a preexisting DNA chain, for which a free 3'-OH in the DNA template is required. DNA polymerase I also can hydrolyze DNA progressively from the 3'-OH terminus, but the nucleotide removed must not be part of a double helix. This activity probably has an editing function during replication. The enzyme also has 5'-3' nuclease activity. The cleaved bond must be in a double-helical region, and the active site is quite separate from the active sites for DNA synthesis and 3'-5' nuclease activity.

210. The answer is C (2, 4). *(Stryer, p 687.)* The inhibitory effect of glucose on β-galactosidase synthesis is an example of a general positive control system that overrides the transcriptional control of many operons related to sugar metabolism. The controlling metabolite is adenosine 3',5'-monophosphate (cyclic AMP), which decreases in concentration when glucose is present. Cyclic AMP action is mediated through the catabolite activator protein (CAP), and transcription of *lac* and *ara* (and some other operons) may occur when the cyclic AMP-CAP complex is bound to the promoter.

211. The answer is C (2, 4). *(Stryer, pp 681-682.)* Operator genes are the loci for binding of repressor or inducer molecules. According to whether or not the operator gene locus is occupied, the structural genes either will or will not be transcribed.

212. The answer is E (all). *(Lehninger, ed 2. pp 959-963.)* Many messenger RNA molecules code for more than one polypeptide chain; this is often the basis for coordinate control of enzyme synthesis in bacteria. There are no "spacers" to mark the end of one codon and the beginning of another. There are three nonsense codons. According to the wobble hypothesis of Crick, certain anti-codons can pair with more than one codon where variability occurs at the 3'-position of the codon.

213. The answer is C (2, 4). *(Mahler, ed 2. pp 296-297.)* RNA is synthesized with only one of the four ribonucleotides, labeled at the alpha position with ^{32}P, for "nearest neighbor" analysis. Since alkaline hydrolysis gives 2'- and 3'-nucleoside monophosphates as products, alkaline hydrolysis allows determination of the frequency with which the 3'-hydroxyl groups of various nucleotides are attached via phosphodiester bonds to the 5'-hydroxyl group of the labeled precursor.

214. The answer is B (1, 3). *(Mahler, ed 2. pp 202-209.)* The correct Watson-Crick base pairings for most DNA molecules are adenine-thymine and cytosine-guanine. In RNA, uracil replaces thymine. Some DNAs, however, depart from the characteristic Chargaff base-pairing, like coliphage ϕX174 whose structure bears a similarity to certain RNAs.

215. The answer is E (all). *(Stryer, pp 625-637.)* The triplet genetic code is degenerate, which is to say that, for most amino acids, there is more than one code word. The triplets of bases (codons) that specify the same amino acid usually differ only in the last base of the triplet. Chain termination is determined by three codons: UAA, UAG, and UGA.

216. The answer is A (1, 2, 3). *(Davis, ed 2. pp 175-176. Hayes, ed 2. pp 56-60.)* Cell fusion allows for genetic recombination and is part of fertilization and conjugation; it proceeds by various mechanisms in different forms of life. In haploid bacteria (such as *E. coli*), cell fusion provides a means of genetic transfer between male and female organisms. In diploid organisms, meiosis yields haploid ova and sperm that fuse by a more complex mechanism.

217. The answer is D (4). *(White, ed 6. pp 817-819.)* In the course of protein synthesis, amino acids are first activated by the formation of enzyme-bound aminoacyl-adenylate complexes from which the aminoacyl group is transferred to tRNA. Both reactions are catalyzed by aminoacyl-tRNA synthetases, which are highly specific for both the amino acid and the tRNA. The peptide chain grows by stepwise additions of residues at its carboxyl end; i.e., it is synthesized from its amino terminus. The movement of the mRNA in relation to the ribosome requires GTP and a protein factor.

218-224. The answers are: 218-A, 219-C, 220-D, 221-B, 222-A, 223-D, 224-B. *(Thorn, ed 8. pp 321-327.)* There are a number of inherited diseases in which the primary defect has been localized to a mutation involving a single protein, or enzyme. In the majority of cases the inheritance is recessive, either autosomal or X-linked. Because a single "dose" of defective gene product is seldom expressed as an impaired phenotype, dominant inheritance is rarer and usually more complicated than recessive inheritance.

In the examples listed in the question, phenylketonuria (lack of phenylalanine hydroxylase) and Wilson's disease (lack of the copper transport protein ceruloplasmin) are transmitted as autosomal recessive disorders. Lesch-Nyhan syndrome (lack of hypoxanthine-guanine phosphoribosyl transferase) and the condition of red-green color blindness are two examples of X-linked recessive inheritance. Autosomal dominant inheritance is seen in acute intermittent porphyria (an abnormal enzyme catalyzing a rate-limiting step in heme synthesis) and also in familial hypercholesterolemia (abnormal cell-surface receptor that binds plasma low-density lipoproteins). Vitamin D-resistant rickets is usually transmitted as an X-linked dominant characteristic, particularly when hypophosphatemia is present

225-228. The answers are: 225-E, 226-D, 227-A, 228-B. *(Mahler, ed 2. pp 914-949).* Messenger RNA (mRNA) is complementary in base sequence to the DNA in an operon and codes for the polypeptide chains that correspond to the genes of the operon. Transcription of DNA into the base sequence of mRNA occurs in the nucleus, but translation of the message in mRNA into a peptide sequence occurs on ribosomes in the cytosol. Transfer RNA (tRNA) carries the anticodon triplet that is complementary to the triplet sequence in mRNA, and is also covalently linked to the amino acid that specifically corresponds to that anticodon. 16-S and 23-S rRNAs form parts of the 30-S and 50-S ribosomal subunits respectively. The 30-S ribosomal subunit binds mRNA, forming an initiation complex for the process of peptide synthesis. The 50-S ribosomal subunit interacts with the 30-S ribosomal subunit and carries the peptidyl transferase enzyme activity which catalyzes the formation of the peptide bond.

Carbohydrates and Lipids

DIRECTIONS: Each question below contains five suggested answers. Choose the **one best** response to each question.

229. Which of the following enzymes functions in both glycolysis and gluconeogenesis?

(A) Pyruvate kinase
(B) Pyruvate carboxylase
(C) Glyceraldehyde 3-phosphate dehydrogenase
(D) Fructose 1,6-diphosphatase
(E) Hexokinase

230. The structure shown below is that of

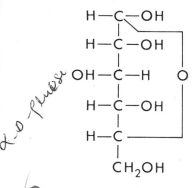

(A) α-D-glucopyranose
(B) β-D-glucopyranose
(C) α-D-glucofuranose
(D) β-L-glucofuranose
(E) α-D-fructofuranose

231. Gluconeogenesis is a process in carbohydrate metabolism which forms

(A) glucose
(B) maltose
(C) sucrose
(D) fructose
(E) glucose 1-phosphate

232. The formation of carbohydrates from amino acids is known as

(A) glycolysis
(B) glycogenolysis
(C) glycogenesis
(D) gluconeogenesis
(E) none of the above

233. Hydrolysis of maltose will yield

(A) glucose only
(B) fructose + glucose
(C) galactose + glucose
(D) mannose + glucose
(E) fructose + galactose

234. The first product of glycogenolysis is

(A) glucose 6-phosphate
(B) glucose 1,6-diphosphate
(C) glucose 1-phosphate
(D) fructose 1-phosphate
(E) glucose

235. The substrate for aldolase is

(A) glucose 6-phosphate
(B) fructose 6-phosphate
(C) fructose 1,6-diphosphate
(D) phosphoglyceric acid
(E) 1,3-diphosphoglyceric acid

236. The allosteric activator of glycogen synthase D is

(A) UTP
(B) ADP
(C) 5'-AMP
(D) glucose 6-phosphate
(E) glucose 1-phosphate

237. Which of the following fatty acids is a biosynthetic precursor of prostaglandin $F_{1\alpha}$?

(A) Hexadecanoic acid
(B) Octadecanoic acid
(C) cis-9-Octadecanoic acid
(D) Eicosa-8,11,14-trienoic acid
(E) Eicosa-4,8,11,14-tetraenoic acid

238. An animal deficient in essential fatty acids would accumulate which of the following polyunsaturated fatty acids?

(A) Docosahexenoic acid
(B) Arachidonic acid
(C) Eicosa-11,14-dienoic acid
(D) γ-Linolenic acid
(E) Eicosa-5,8,11-trienoic acid

239. Which of the following fatty acids can be synthesized de novo in higher animals?

(A) Eicosa-5,8,11-trienoic acid
(B) Arachidonic acid
(C) Docosahexenoic acid
(D) Linolenic acid
(E) Linoleic acid

240. Diglyceride + NuDP-choline ⇒ NuMP + phosphatidylcholine. In the above reaction, NuMP stands for

(A) AMP
(B) CMP
(C) GMP
(D) TMP
(E) UMP

241. The cholesterol molecule is a

(A) benzene derivative
(B) quinoline derivative
(C) steroid
(D) tocopherol
(E) straight chain acid

242. Cholesterol is synthesized from

(A) pyruvate
(B) oxalate
(C) malate
(D) acetyl CoA
(E) α-ketoglutaric acid

243. Which of the following steps in the biosynthesis of cholesterol is thought to be rate-controlling and the locus of metabolic regulation?

(A) Geranyl pyrophosphate → farnesyl pyrophosphate
(B) Squalene → lanosterol
(C) Lanosterol → cholesterol
(D) 3-Hydroxy-3-methylglutaryl CoA → mevalonic acid
(E) None of the above

244. Which of the following sterols is formed directly by the cyclization of squalene in mammals?

(A) Cholesterol
(B) Demosterol
(C) Lanosterol
(D) β-Sitosterol
(E) Cortisol

245. Cholesterol is a precursor of

(A) coenzyme A
(B) ubiquinone
(C) vitamin A
(D) vitamin D
(E) vitamin E

246. The "committed step" in the biosynthesis of cholesterol from acetyl CoA is catalyzed by which of the following enzymes?

(A) 3-Hydroxy-3-methylglutaryl CoA synthetase
(B) 3-Hydroxy-3-methylglutaryl CoA reductase
(C) 3-Hydroxy-3-methylglutaryl cleavage enzyme
(D) Mevalonate kinase
(E) Squalene epoxidase

247. Which of the following is a phospholipid?

(A) Glycogen
(B) Sphingomyelin
(C) Prostaglandin
(D) Oleic acid
(E) Triglyceride

248. All the following compounds are phospholipids EXCEPT

(A) cephalins
(B) cerebrosides
(C) lecithins
(D) sphingomyelins
(E) plasmalogens

249. The phospholipid cardiolipin is found almost exclusively in

(A) mitochondrial membranes
(B) plasma membranes
(C) lysosome membranes
(D) smooth endoplasmic reticular membranes
(E) rough endoplasmic reticular membranes

250. Which of the following is NOT a constituent of ganglioside molecules?

(A) Glycerol
(B) Sialic acid
(C) Hexose sugar
(D) Sphingosine
(E) Long-chain fatty acid

251. Which of the following compounds has the lowest density?

(A) Chylomicrons
(B) β-Lipoproteins
(C) Pre-β-lipoproteins
(D) α-Lipoproteins
(E) Transferrin

252. Cyanides produce hypoxia by

(A) producing central hypoventilation
(B) interfering with oxygen carriage
(C) slowing capillary circulation
(D) inhibiting cellular respiration
(E) none of the above mechanisms

253. Which of the following statements most accurately describes glucose uptake from the intestinal lumen?

(A) Glucose crosses the mucosal membrane by a process of passive diffusion down a concentration gradient
(B) Glucose crosses the mucosal membrane by a carrier-mediated transport process requiring the concomitant transfer of Na^+ in the same direction
(C) Glucose crosses the serosal membrane by a carrier-mediated transport process requiring the concomitant transfer of Na^+ in the opposite direction
(D) Glucose is actively transported across the serosal membrane against a concentration gradient
(E) Glucose uptake is independent of high-energy phosphate bonds

254. The inborn error of metabolism known as Refsum's disease involves an accumulation of high concentrations of phytanic acid (3,7,11,15-tetramethyl hexadecanoic acid) in the tissues and serum. This condition results from an impairment in which of the following enzymatic pathways?

(A) β-Oxidation of fatty acids
(B) α-Oxidation of fatty acids
(C) ω-Oxidation of fatty acids
(D) The formation of acyl carnitines
(E) Conversion of propionyl CoA to succinyl CoA

255. The number of net molecules of ATP yielded in the conversion of one glucosyl residue in glycogen to two molecules of lactate is

(A) one
(B) two
(C) three
(D) four
(E) five

256. The ratio that most closely approximates the number of net molecules of ATP formed per mole of glucose utilized under aerobic conditions to the net number formed under anaerobic conditions is

(A) 2:1
(B) 9:1
(C) 13:1
(D) 18:1
(E) 25:1

257. The Pasteur effect refers to

(A) an increase in hexokinase activity due to increased oxidative phosphorylation
(B) an increase in ethanol formation from pyruvate upon changing from an anaerobic to aerobic metabolism
(C) an increase in glucose utilization via the pentose phosphate pathway upon changing from an anaerobic to aerobic metabolism
(D) a decrease in glucose utilization upon changing from an anaerobic to aerobic metabolism
(E) a decrease in the respiratory quotient upon changing from carbohydrate to fat as the major metabolic fuel

258. Which of the following enzymes of the glycolytic pathway is particularly sensitive to inhibition by fluoride ions?

(A) Hexokinase
(B) Aldolase
(C) Pyruvate kinase
(D) Enolase
(E) Phosphohexose isomerase

259. The net number of ATP molecules formed per molecule of glucose in aerobic glycolysis is

(A) 2
(B) 6
(C) 18
(D) 36
(E) 54

260. Which of the following energy-related activities does NOT occur in mitochondria?

(A) Citric acid cycle
(B) Fatty acid oxidation
(C) Electron transport
(D) Glycolysis
(E) Oxidative phosphorylation

261. All the following compounds contain a high-energy phosphate bond EXCEPT

(A) ADP
(B) creatine phosphate
(C) glucose 6-phosphate
(D) phosphoenolpyruvate
(E) 1,3-diphosphoglycerate

262. In the diagram of ADP shown below, four bonds are labeled A through D. Which is a high-energy bond?

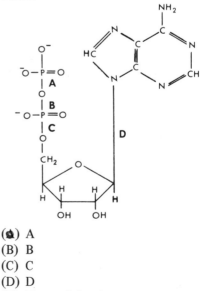

(A) A
(B) B
(C) C
(D) D
(E) None of the above

263. In the figure shown below, fructose 1,6-diphosphate is located at point

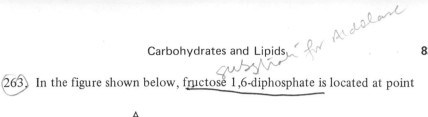

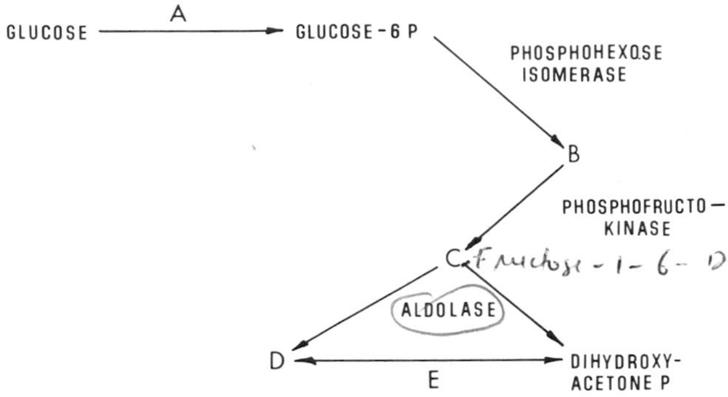

GLUCOSE —A→ GLUCOSE-6 P

PHOSPHOHEXOSE ISOMERASE

B

PHOSPHOFRUCTO-KINASE

C *Fructose - 1- 6- Diphosphate.*

ALDOLASE

D ←—E—→ DIHYDROXY-ACETONE P

(A) A
(B) B
(●) C
(D) D
(E) E

264. Among the many molecules of high-energy phosphate compounds formed as a result of the functioning of the tricarboxylic acid cycle, one molecule is synthesized at the substrate level. In which of the following steps does this occur?

(A) Citrate → α-ketoglutarate
(●) α-Ketoglutarate → succinate
(C) Succinate → fumarate
(D) Fumarate → malate
(E) Malate → oxaloacetate

265. The process of glycolysis includes the following reactions and their concomitant free-energy changes:

Glyceraldehyde 3-phosphate + NAD$^+$ + P$_i$ $\rightleftharpoons$ 1,3-diphosphoglycerate + NADH + H$^+$: $\Delta G^{o\prime}$ = +1.5 kcal/mol

1,3-Diphosphoglycerate + ADP $\rightleftharpoons$ 3-phosphoglycerate + ATP: $\Delta G^{o\prime}$ = -4.5 kcal/mol

For the two-step process converting glyceraldehyde 3-phosphate to 3-phosphoglycerate, the overall free energy change is

(A) $\Delta G^{o\prime}$ = +6.0 kcal/mol
(B) $\Delta G^{o\prime}$ = +3.0 kcal/mol
(●) $\Delta G^{o\prime}$ = -3.0 kcal/mol
(D) $\Delta G^{o\prime}$ = -4.5 kcal/mol
(E) $\Delta G^{o\prime}$ = -6.0 kcal/mol

266. The fructokinase reaction produces which of the following intermediates?

(A) Fructose 1-phosphate
(B) Fructose 6-phosphate
(C) Fructose 1,6-diphosphate
(D) Glyceraldehyde and dihydroxy-acetone phosphate
(E) Glyceraldehyde 3-phosphate and dihydroxyacetone phosphate

267. The diagram of the citric acid cycle, shown below, contains lettered steps where H^+- e^- pairs might be given to the electron transport chain. At which step does donation of an H^+- e^- pair NOT occur?

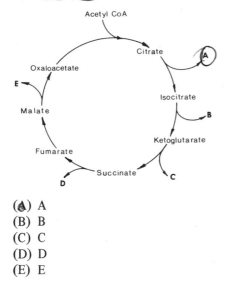

Acetyl CoA
Citrate
Oxaloacetate
Isocitrate
Malate
Ketoglutarate
Fumarate
Succinate

(A) A
(B) B
(C) C
(D) D
(E) E

268. All the following enzymes are involved in the citric acid (Krebs) cycle EXCEPT

(A) fumarase
(B) isocitrate dehydrogenase
(C) succinate thiokinase
(D) pyruvate dehydrogenase
(E) aconitase

269. Living cells can use all the following sources of energy for their metabolic function EXCEPT

(A) ATP
(B) fats
(C) sugars
(D) ambient heat
(E) sunlight

270. Which of the following statements about the citric acid cycle is true?

(A) It contains no intermediates for glucogenesis
(B) It contains intermediates for amino acid synthesis
(C) It generates fewer molecules of ATP than glycolysis, per mole of glucose consumed
(D) It is an anaerobic process
(E) It is the major anabolic pathway for glucose synthesis

271. An allosteric enzyme thought to be responsible for controlling the rate of the citric acid cycle is

(A) pyruvate dehydrogenase
(B) aconitase
(C) isocitrate dehydrogenase
(D) malate dehydrogenase
(E) citrate dehydrogenase

272. The monosaccharide most rapidly absorbed from the small intestines is

(A) xylose
(B) glucose
(C) fructose
(D) mannose
(E) galactose

273. Which of the following comparisons between hexokinase and glucokinase is FALSE?

(A) The Michaelis constant of hexokinase for glucose is much smaller than that of glucokinase
(B) Hexokinase is less specific about sugars it will accept as substrate than is glucokinase
(C) Only hexokinase is inhibited by glucose 6-phosphate
(D) Only glucokinase is present in the brain
(E) Hexokinase and glucokinase are present in liver tissue

274. All the following carbohydrates are disaccharides EXCEPT

(A) cellobiose
(B) cellulose
(C) lactose
(D) maltose
(E) sucrose

275. The Cori cycle may be described as

(A) the interconversion between glycogen and glucose 1-phosphate
(B) the synthesis of alanine from pyruvate in skeletal muscle and the synthesis of pyruvate from alanine in liver
(C) the synthesis of urea in liver and degradation of urea to carbon dioxide and ammonia by bacteria in the gut
(D) the production of lactate from glucose in peripheral tissues with the resynthesis of glucose from lactate in liver
(E) none of the above

276. One of the principal sources of the hydrogen stored in the form of NADPH is

(A) glycolysis
(B) oxidative phosphorylation
(C) the synthesis of fatty acids
(D) the citric acid cycle
(E) the hexose monophosphate shunt

277. How many ATP molecules are required to convert two molecules of lactate into glucose in mammalian liver?

(A) Two
(B) Three
(C) Four
(D) Five
(E) Six

278. Gluconeogenesis refers to the intracellular synthesis of "new" glucose from noncarbohydrate precursors. All the following compounds will yield a net synthesis of glucose EXCEPT

(A) aspartic acid
(B) glutamic acid
(C) succinic acid
(D) leucine
(E) phosphoenolpyruvate

279. In the reaction below, NuDP stands for

NuDP glucose + glycogen$_n$ →
NuDP + glycogen$_{n+1}$

(A) ADP
(B) CDP
(C) GDP
(D) TDP
(E) UDP

280. In the reaction below, NuTP stands for

NuTP + glucose → glucose 6-phosphate + NuDP

(A) ATP
(B) CTP
(C) GTP
(D) TTP
(E) UTP

281. In the reaction below, NuTP stands for

oxaloacetate + NuTP → NuDP
+ phosphoenolpyruvate + CO_2

(A) ATP
(B) CTP
(C) GTP
(D) TTP
(E) UTP

Questions 282-283

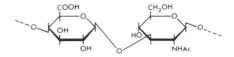

282. The structure shown above is the mucopolysaccharide

(A) chitin
(B) chondroitin sulfate
(C) heparin
(D) hyaluronic acid
(E) keratin

283. The mucopolysaccharide shown above is found primarily in the

(A) cartilage
(B) hard, outer shells of insects
(C) mast cells
(D) outer layers of the epidermis
(E) vitreous humor of the eye

284. The active form of sulfate used in the biosynthesis of sulfate ester is which of the following?

(A) Chondroitin sulfate
(B) 3'-Phosphoadenosine 5'-phosphosulfate
(C) Adenosine 5'-phosphosulfate
(D) Sulfonyl urea
(E) None of the above

285. Hydrolysis of a fat by an alkali is given the special name of

(A) esterification
(B) reduction
(C) saponification
(D) oxidation
(E) hydrolysis

286. The *de novo* biosynthesis of fatty acids

(A) does not utilize acetyl CoA
(B) produces only fatty acids shorter than ten carbon atoms
(C) requires the intermediate malonyl CoA
(D) takes place primarily in mitochondria
(E) uses NAD as an oxidizing agent

287. In the pathway leading to biosynthesis of acetoacetate from acetyl CoA in liver, the immediate precursor of acetoacetate is which of the following?

(A) 3-Hydroxybutyrate
(B) Acetoacetyl CoA
(C) 3-Hydroxybutyryl CoA
(D) Mevalonic acid
(E) 3-Hydroxy-3-methylglutaryl CoA

288. The rate-limiting step in the extramitochondrial synthesis of fatty acids is

(A) condensing enzyme
(B) hydratase
(C) acetyl CoA carboxylase
(D) acyl transferase
(E) palmitoyl deacylase

289. Continuation of the fatty acid oxidation cycle in the metabolism of long-chain fatty acids is dependent on the presence of all the following enzymes EXCEPT

(A) acyl CoA dehydrogenase
(B) β-hydroxyacyl CoA dehydrogenase
(C) enoyl hydrase
(D) β-ketothiolase
(E) thiokinase

290. In the biosynthesis of fatty acids, the compound that transports the acetate groups out of the mitochondria into the cytoplasm is which of the following?

(A) Acetyl CoA
(B) Acetyl carnitine
(C) Acetyl phosphate
(D) Citrate
(E) None of the above

291. The oxidation and degradation of fatty acids in the cell

(A) begins with the fatty acid thioester of CoA
(B) does not produce useful energy for the cell
(C) is inhibited by carnitine
(D) occurs primarily in the nucleus
(E) proceeds through successive shortening of the fatty acids by three-carbon units

292. In the biosynthesis of triglycerides from glycerol 3-phosphate and acyl CoAs, the first intermediate formed is

(A) 2-monoacyl glycerol
(B) 1,2-diacyl glycerol
(C) lysophosphatidic acid
(D) phosphatidic acid
(E) acyl carnitine

DIRECTIONS: Each question below contains four suggested answers of which **one** or **more** is correct. Choose the answer:

A	if	**1, 2, and 3**	are correct
B	if	**1 and 3**	are correct
C	if	**2 and 4**	are correct
D	if	**4**	is correct
E	if	**1, 2, 3, and 4**	are correct

293. Oligomycin is thought to interfere with synthesis of energy-rich compounds, like ATP, by

(1) dissociating cytochrome c from mitochondrial membranes
(2) uncoupling electron transfer between NAD and flavoprotein
(3) blocking the carnitine shuttle in mitochondria
(4) inhibiting mitochondrial ATPase

294. Which of the following sugars would be expected to form the same osazone as glucose?

(1) Mannose
(2) Galactose
(3) Fructose
(4) Ribose

295. Which of the following are ketose sugars?

(1) Ribose
(2) Ribulose
(3) Glucose
(4) Fructose

296. Which of the following compounds contain sugar residues?

(1) ATP
(2) NAD
(3) RNA
(4) Acetyl CoA

297. Which of the following are substrates for gluconeogenesis in mammalian liver?

(1) Oleic acid
(2) Serine
(3) Leucine
(4) Glycerol

298. The polysaccharide glycogen is

(1) a structural support for liver cell membranes
(2) a copolymer of glucose and galactose
(3) branched at few points
(4) hydrolyzed by α-1, 4-glucan maltohydrolase to maltose

299. Which of the following statements about the structure of glycogen are true?

(1) There are α-1,4 glycosidic linkages
A (2) There are α-1,6 glycosidic linkages
(3) All the monosaccharides in glycogen are α-D-glucose
(4) Glycogen is an unbranched molecule

300. The irreversible enzymes bypassed in gluconeogenesis include

(1) phosphofructokinase
A (2) hexokinase
(3) pyruvate kinase
(4) enolase

UDP-GGG-T
301. Glycogen synthetase, the enzyme involved in the biosynthesis of glycogen, may

(1) be more specifically defined as a UDP-glucose-glycogen glycosyl transferase
B (2) be able to synthesize glycogen without a polymer primer
(3) exist in active and inactive forms, subject in part to hormonal control of activation
(4) employ UDP-D-glucose as a glycosyl donor in both plants and animals

302. The hexose monophosphate shunt includes which of the following enzymes?

(1) Fumarase
(2) α-Ketoglutarate dehydrogenase
(3) Hexokinase
D (4) Glucose 6-phosphate dehydrogenase

303. Which of the following enzymes can be regarded as catalyzing an important "decision-making step" whereby liver cells elect to expend a glucose carbon for energy needs, instead of retaining it in response to blood glucose level requirements?

(1) Glucose 6-phosphate dehydrogenase
(2) Glyceraldehyde 3-phosphate dehydrogenase
(3) Pyruvate carboxylase
D (4) Pyruvate dehydrogenase

304. Oxidative decarboxylation of pyruvate involves

(1) NAD
B (2) NADP
(3) FAD
(4) folic acid

305. The citric acid cycle is inhibited by

(1) arsenite
(2) malonate
(3) fluroacetate
E (4) anaerobic conditions

306. Factors affecting the activity of the citric acid cycle probably include

(1) levels of oxaloacetic acid
(2) levels of NAD^+
(3) ratio of concentrations of ADP/ATP
(4) number of mitochondria per cell

307. The reaction catalyzed by α-ketoglutarate dehydrogenase in the citric acid cycle requires

(1) NAD
(2) NADP
(3) CoA
(4) ATP

308. Glycolysis in the Embden-Meyerhof pathway is typified by

(1) the anaerobic conversion of glucose to lactic acid in mammalian muscle
(2) a requirement for O_2, in order for yeast to convert glucose to CO_2 and ethanol
(3) the ability to proceed independent of P_{O_2}
(4) a net gain of one mole of ATP per mole of glucose traversing the pathway under aerobic conditions

309. Which of the following are inhibitors of the enzyme phosphofructokinase?

(1) Citrate
(2) Cyclic AMP
(3) ATP
(4) NH_4^+

310. Which of the following statements are consistent with the chemiosmotic hypothesis of Mitchell (a currently popular model to explain mitochondrial oxidative phosphorylation)?

(1) The energy for ATP synthesis is derived from re-entry of protons into the mitochondrion down an electrochemical gradient
(2) Uncouplers (dinitrophenol) can increase the permeability of artificial lipid membranes to protons
(3) As electrons are passed down the electron transport chain, there is a separation of charge which results in active transport out of mitochondria
(4) The pH is ordinarily lower inside the mitochondrion than outside

311. Which of the following statements concerning "energy-rich" compounds of biologic systems are true?

(1) Only phosphate esters serve as "energy-rich" compounds
(2) Amino acid esters have free energies of hydrolysis comparable to that of ATP
(3) ATP, an "energy-rich" compound, has a positive free energy of hydrolysis
(4) The nature of "energy-rich" compounds is explained on the basis of resonance theory

SUMMARY OF DIRECTIONS

A	B	C	D	E
1, 2, 3	1, 3	2, 4	4	All are
only	only	only	only	correct

312. Which of the following statements describe the mechanism by which reducing equivalents from NADH are thought to be transported into mitochondria from the cytoplasm for the purpose of aerobic oxidation?

(1) NADH is transported across the mitochondrial membranes directly

(2) Dihydroxyacetone phosphate is reduced to glycerol 3-phosphate which in turn is oxidized by a flavoprotein in the inner membrane of the mitochondrion

(3) Oxaloacetate is reduced to malate which enters the mitochondrion where it is oxidized to oxaloacetate which then exits from the mitochondrion

(4) Oxaloacetate is reduced to malate which enters the mitochondrion where it is oxidized to oxaloacetate and then transaminated to aspartate for transport out of the mitochondrion

313. Which of the following statements about skeletal muscle glycogen phosphorylase are correct?

(1) Cyclic AMP leads indirectly to the conversion of glycogen phosphorylase to the active, phosphorylated form

(2) Glycogen phosphorylase is the enzyme responsible for both biosynthesis and degradation of glycogen

(3) AMP is an allosteric activator of the inactive form of the enzyme phosphorylase *b*

(4) Exposure of muscle to the hormone insulin leads to activation of glycogen phosphorylase

314. High-energy phosphate bonds are found in

(1) phosphoenolpyruvate
(2) ATP
(3) creatine phosphate
(4) ADP

315. Which of the following are examples of biologically important mucopolysaccharides?

(1) Neuraminidase
(2) Hyaluronic acid
(3) Cellulose
(4) Heparin

316. Which of the following statements about the electron transport chain are true?

(1) Continuation of the process is dependent on oxidative phosphorylation
(2) Three moles of ATP are formed during the transfer of electrons from NAD to oxygen
(3) Free-energy change is positive for the transfer of electrons from NADH to oxygen
(4) The iron ion in the poryphyrin ring undergoes $Fe^{++} \rightleftharpoons Fe^{+++}$ transitions

317. Which of the following statements about polysaccharides are true?

(1) They are a major source of biologic energy
(2) They exist in either linear or branched forms
(3) They are important structural elements in bacterial cell walls
(4) They are informational molecules

318. Compounds normally used by the body to conjugate bile acids include

(1) glycine
(2) glucuronic acid
(3) taurine
(4) fatty acids

319. Which of the following statements about lipids are true?

(1) They are an intracellular energy source
(2) They are poorly soluble in water
(3) They are structural components of membranes
(4) They are composed of only carbon, hydrogen, and oxygen

320. In the pathway below, ATP is produced between

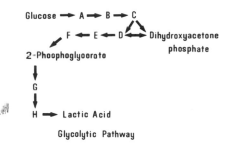

Glycolytic Pathway

(1) A-B
(2) F and 2-phosphoglycerate
(3) C-D
(4) G-H

321. Which of the following statements about ketone bodies are true?

(1) They include acetone and acetoacetic acid
(2) They may be excreted in urine
(3) They may result from starvation
(4) They are related to diabetes insipidus

SUMMARY OF DIRECTIONS

A	B	C	D	E
1, 2, 3 only	1, 3 only	2, 4 only	4 only	All are correct

322. Important intermediates in the synthesis of fatty acids from glucose in animal tissues include

(1) carnitine
(2) pyruvate
(3) ATP
(4) acetyl CoA

_c_

323. Which of the following fatty acids are dietary essentials in humans?

(1) Palmitic acid
(2) Stearic acid
(3) Oleic acid
(4) Linolenic acid

_D_

324. Which of the following statements are true of the synthesis of fatty acids from acetyl CoA but not of the oxidation of fatty acids to acetyl CoA?

(1) All oxidation-reduction steps use NADPH as a cofactor
(2) CoA is the only pantetheine-containing substance involved in the pathway
(3) Malonyl CoA is an "activated" intermediate
(4) The reactions proceed within mitochondria

_B_

325. Which of the following are intermediates in the metabolism of propionic acid?

(1) Propionyl CoA
(2) D-Methylmalonyl CoA
(3) L-Methylmalonyl CoA
(4) Succinyl CoA

_E_

326. Ethanol is converted in the liver to

(1) acetone
(2) acetaldehyde
(3) methanol
(4) acetyl CoA

_c_

327. Biosynthetic precursors of cholesterol include

(1) lanosterol
(2) mevalonic acid
(3) squalene
(4) progesterone

_A_

328. An 11β-hydroxylase is required in the biosynthesis of

(1) estradiol
(2) cortisol
(3) testosterone
(4) aldosterone

_c_

329. A 17-hydroxylase is required in the biosynthesis of

(1) estradiol
(2) cortisol
(3) testosterone
(4) aldosterone

_A_

330. A 21 hydroxylase enzyme is involved in the biosynthesis of

(1) estradiol
(2) cortisol
(3) testosterone
(4) aldosterone

331. Hydrolysis of a mixture of phosphoglycerides will yield

(1) choline
(2) glycerol
(3) phosphate
(4) serine

332. Chylomicrons are composed of

(1) triglycerides
(2) cholesterol
(3) phospholipids
(4) protein

333. Which of the following statements about the glyoxylate cycle are true?

(1) It is found in microorganisms as well as in higher forms of life
(2) It oxidizes acetate completely to CO_2
(3) It functions as an alternate to the citric acid cycle in bacteria
(4) It is essential to microorganisms in media whose only carbon source is acetate

334. The presence of starch granules indicates

(1) an aldose-ketose transformation
(2) activation of the Calvin cycle ("dark reaction")
(3) that a prokaryote has metabolized glucose
(4) activation of the pentose phosphate pathway

335. Levan is which of the following?

(1) A polymer of D-fructose
(2) Analogous to dextran
(3) Formed from sucrose by levansucrase
(4) A secondary metabolite of *Bacillus brevis*

336. Dextran is which of the following?

(1) A long-chain polymer of sucrose
(2) Forms part of dental plaque
(3) A long-chain fructose polymer
(4) A long-chain glucose polymer

337. Lipid A may correctly be said to

(1) form part of the plasma membrane
(2) form part of the ketodeoxyoctonic acid complex in the cell wall
(3) contain the core sugars
(4) be attached to the core polysaccharide

338. The beta-oxidation of fatty acids occurs mainly in the

(1) cytosol
(2) cell membrane
(3) absence of ATP
(4) mitochondria

SUMMARY OF DIRECTIONS

A	B	C	D	E
1, 2, 3 only	1, 3 only	2, 4 only	4 only	All are correct

339. Lipoic acid is essential for

(1) glycolysis
(2) acetyl CoA carboxylase reactions
(3) the transamination of α-keto-glutaric acid
(4) oxidative decarboxylation of pyruvate

340. Which of the following statements concerning the regulatory effects of citrate in liver are correct?

(1) It activates phosphofructokinase
(2) It activates acetyl CoA car-boxylase
(3) It activates enolase
(4) It inhibits pyruvate kinase

341. In animals, the desaturation of essential fatty acids occurs in which of the following subcellular fractions?

(1) Nuclear
(2) Supernatants of 100,000 x g (60 min)
(3) Mitochondrial
(4) Microsomal

Questions 342-344

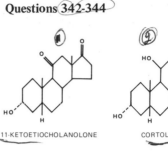

11-KETOETIOCHOLANOLONE

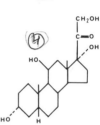

CORTOL

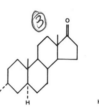

ANDROSTERONE   TETRAHYDROCORTISOL

342. Which of the following are 17-ketosteroids?

(1) 11-Ketoetiocholanolone
(2) Cortol
(3) Androsterone
(4) Tetrahydrocortisol

343. Which of the following compounds give a positive Porter-Silber color reaction, i.e., are "17-hydroxy-steroids?"

(1) 11-Ketoetiocholanolone
(2) Cortol
(3) Androsterone
(4) Tetrahydrocortisol

344. Which of the following are 17-ketogenic steroids?

(1) 11-Ketoetiocholanolone
(2) Cortol
(3) Androsterone
(4) Tetrahydrocortisol

DIRECTIONS: The groups of questions below consist of lettered choices followed by several numbered items. For each numbered item select the **one** lettered choice with which it is **most** closely associated. Each lettered choice may be used once, more than once, or not at all.

Questions 345-347

For each disaccharide that follows, choose the monosaccharide(s) of which it is composed.

    (A) Fructose
    (B) Glucose
    (C) Glucose and fructose
    (D) Glucose and galactose
    (E) Fructose and galactose

C 345. Sucrose

D 346. Lactose

B 347. Maltose

Questions 348-349

For each description that follows, choose the structural carbohydrate with which it is most likely to be associated.

    (A) Keratan sulfate
    (B) Teichoic acid
    (C) Starch
    (D) Cellulose
    (E) Murein

E 348. Polysaccharide component of bacterial cell walls

A 349. Principal, final polysaccharide component of connective tissue

Questions 350-352

For each reaction that follows, select the lettered step in the glycolytic sequence with which it is most closely related.

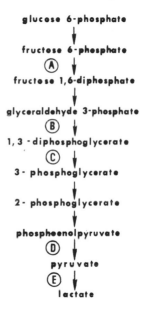

glucose 6-phosphate
↓
fructose 6-phosphate
(A) ↓
fructose 1,6-diphosphate
↓
glyceraldehyde 3-phosphate
(B) ↓
1,3 - diphosphoglycerate
(C) ↓
3- phosphoglycerate
↓
2- phosphoglycerate
↓
phosphoenolpyruvate
(D) ↓
pyruvate
(E) ↓
lactate

A 350. ATP → ADP

E 351. NADH → NAD

B 352. NAD → NADH

Questions 353-355

For each of the following structures accompanied by its physiologic actions, choose the compound with which it is most likely to be identified.

(A) Prostaglandin A
(B) Prostaglandin E
(C) Prostaglandin F
(D) Thyroid-stimulating hormone
(E) Follicle-stimulating hormone

STRUCTURE	EFFECT ON BLOOD PRESSURE	NONVASCULAR SMOOTH MUSCLE STIMULATING ACTIVITY
A 353.	DECREASE	INACTIVE
B 354. 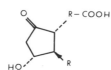	DECREASE	VERY ACTIVE
C 355. 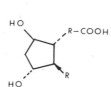	TRANSIENT INCREASE	VERY ACTIVE

Carbohydrates and Lipids

Answers

229. The answer is C. *(Stryer, p 369.)* Glyceraldehyde 3-phosphate dehydrogenase functions in both glycolysis and gluconeogenesis. Pyruvate carboxylase and fructose 1,6-diphosphatase are gluconeogenic enzymes only, whereas pyruvate kinase and hexokinase are glycolytic enzymes.

230. The answer is A. *(Stryer, pp 279-281, 304-305.)* The molecule depicted in the question is α-D-glucose. It is one of a series of D-glucose hemiacetals in which there are alternate α and β forms available because of the asymmetry in the terminal carbon. Glycosides, which are formed from the β form, are of pharmacologic significance, as in the steroid-containing cardiac glycosides, for example.

231. The answer is A. *(White, ed 6. pp 457-460.)* Gluconeogenesis is defined as the biosynthesis of free glucose, and applies to the aerobic resynthesis of glucose from lactic acid, and other metabolic intermediates, indirectly. It is the reverse of glycolysis, although the two reactions utilize different sets of enzymes.

232. The answer is D. *(Stryer, pp 367-368.)* The formation of glucose from noncarbohydrate precursors such as lactate, glycerol, and amino acids, is known as gluconeogenesis. Glycolysis is the process whereby glucose (or glycogen) is broken down to pyruvate and lactic acid. Glycogenolysis and glycogenesis refer to glycogen breakdown and synthesis respectively.

233. The answer is A. *(Stryer, pp 397-400.)* Hydrolysis of maltose yields the monosaccharide glucose as the only product, as maltose comprises a disaccharide of α-D-glucopyranose. Glucose is combined with fructose in sucrose, and with galactose in lactose.

234. The answer is C. *(Stryer, pp 380-381.)* In glycogenolysis, phosphorylase catalyzes the synthesis of glucose 1-phosphate from glycogen and inorganic phosphate. Phosphoglucomutase converts glucose 1-phosphate to glucose 6-phosphate. Glucose 6-phosphate is a central compound at the junction of glycolytic, gluconeogenic, hexose monophosphate shunt, and glycogenic pathways. Glucose 6-phosphate may give rise to fructose 6-phosphate, and the Embem-Meyerhof pathway of glycolysis.

235. The answer is C. *(Lehninger, ed 2. pp 425-426.)* Aldolase catalyzes the conversion of D-fructose 1, 6-diphosphate to dihydroxyacetone phosphate and D-glyceraldehyde 3-phosphate. Class I fructose diphosphate aldolases, found in higher animals, appear in various isoenzyme forms. In the rabbit, aldolase A occurs primarily in muscle, aldolase B in liver, and aldolase C in brain. The isozymic forms contain polypeptide subunits that vary in amino acid composition.

236. The answer is D. *(Lehninger, ed 2. pp 646-647.)* The dependent form of glycogen synthase (glycogen synthase D) is an allosteric enzyme that is activated by glucose 6-phosphate. A protein phosphatase (glycogen synthase phosphatase) converts glycogen synthase D to the nonallosteric I form, which is active in the absence of glucose 6-phosphate.

237. The answer is D. *(White, ed 6. pp 40, 637.)* Eicosa-8,11,14-trienoic acid and arachidonic (eicosa-5,8,11,14-tetraenoic) acid are the 20-carbon fatty acid precursors of the PG_1 and PG_2 families of prostaglandins respectively. Thus, prostaglandin $F_{1\alpha}$ arises from eicosatrienoic acid. *cis*-9-Octadecanoic acid is the 18-carbon mono-unsaturated fatty acid, oleic; octadecanoid, the 18-carbon saturated fatty acid, stearic; and hexadecanoic acid the 16-carbon saturated fatty acid, palmitic.

238. The answer is E. *(White, ed 6. pp 593-594.)* Eicosa-5,8,11-trienoic acid, being a member of the oleic acid family, can be synthesized *de novo* in mammals. It is this acid that is observed to accumulate in the presence of a deficiency of essential fatty acids, rather than the anabolites of linoleic and linolenic acids, and their family members, comprising the other acids listed in the question.

239. The answer is A. *(White, ed 6. pp 593-594.)* Polyunsaturated fatty acids that mammals are capable of synthesizing *de novo*, such as eicosa-5,8,11-trienoic acid, are characterized by double bonds located exclusively between the seventh carbon of the terminal methyl group and the carboxyl group. Of the remaining possibilities listed in the question, dietary linoleic acid is required for the synthesis of arachidonic acid and dietary linolenic acid for the synthesis of docosa-hexenoic acid. These four polyunsaturated fatty acids all have double bonds within the terminal seven carbon atoms and cannot be made *de novo*.

240. **The answer is B.** *(McGilvery, p 425.)* Choline initially is phosphorylated by a kinase with ATP, and then reacts with cytidine triphosphate (CTP) to yield CDP-choline. CDP-choline is able to react with diglycerides to yield phosphatidylcholine, the most common phosphatidate, thus:

Diglyceride + CDP-choline → CMP + phosphatidylcholine.

241. **The answer is C.** *(White, ed 6. pp 57-60.)* Cholesterol is a steroid. This class of compounds is derived from cyclopentaphenanthrenes, comprising three fused cyclohexane rings, and a terminal cyclopentane ring. Steroids with 8 to 10 carbon atoms at position 17, and an alcoholic hydroxyl group at position 3, are classed as sterols, of which cholesterol is the chief representative in animal tissues.

242. **The answer is D.** *(White, ed 6. pp 619-623.)* Cholesterol is synthesized from acetyl CoA via 3-hydroxy-3-methylglutaryl CoA, which leads in turn by a NADPH-catalyzed reaction to mevalonic acid. Six 5-carbon units from mevalonic acid undergo condensation, producing squalene, a sterol precursor.

243. **The answer is D.** *(White, ed 6. p 628.)* Regulation of cholesterol metabolism is by definition exerted at the "committed" and rate-controlling step. This is the reaction catalyzed by 3-hydroxy-3-methylglutaryl CoA reductase. Reductase activity is reduced by fasting, and by cholesterol feeding, thus providing effective feedback control of cholesterol metabolism.

244. **The answer is C.** *(White, ed 6. pp 619-623.)* Squalene cyclizes to form lanosterol directly, which in turn is converted to cholesterol, via desmosterol. Squalene cyclization into a tetracyclic steroidal compound employs first an epoxidase and activated molecular oxygen, and then an anaerobic cyclase, which yields lanosterol exclusively in mammals.

245. **The answer is D.** *(Lehninger, ed 2. pp 296-298, 335.)* Cholesterol in the skin is converted by sunlight into cholecalciferol (vitamin D_3) via 7-dehydro-cholesterol. The unrelated vitamin A is a carotenoid, vitamin E an aromatic compound with a side chain similar to that of the terpenes, and ubiquinone a cyclic compound with a long isoprenoid side chain. With the exception of coenzyme A, all the compounds listed in the question can be regarded as being derived from isoprene.

246. The answer is B. *(Lehninger, ed 2. pp 554, 681-683.)* The committed step in cholesterol biosynthesis is the reduction of 3-hydroxy-3-methylglutaryl CoA (HMGCoA) to form mevalonate and CoA. HMGCoA is formed from acetyl CoA and acetoacetyl CoA by HMGCoA synthetase, and HMGCoA cleavage enzyme converts HMGCoA to acetyl CoA and acetoacetate, the precursor of ketone bodies. Mevalonate kinase catalyzes the first phosphorylation of mevalonate and squalene epoxidase forms squalene 2,3-epoxide from squalene.

247. The answer is B. *(White, ed 6. p 52.)* Sphingomyelin, found chiefly in neural cells, is an example of a phospholipid. Glycogen is the form in which carbohydrate is stored in the liver and muscle, while oleic acid is a fatty acid. Prostaglandins, once believed to derive exclusively from the prostate gland, are fatty acid derivatives widely distributed in animal tissue.

248. The answer is B. *(White, ed 6. pp 53-54.)* Cephalins, lecithins, sphingomyelins, and plasmalogens all are phospholipids. Three important glycosphingolipids are the cerebrosides, the gangliosides, and the ceramide oligosaccharides. Cerebrosides, which are most abundant in the myelin sheath of nerves, are glycolipids that lack phosphorous.

249. The answer is A. *(White, ed 6. pp 343-344.)* Cardiolipin is found almost exclusively in mitochondrial membranes. Although it is antigenic, antibodies prepared against isolated cardiolipins fail to react with intact membrane preparations from cells, suggesting that cardiolipin is buried within the matrix of the membrane.

250. The answer is A. *(Mahler, ed 2. p 736. White, ed 6. p 616.)* Gangliosides appear to play special roles in the cell plasma membrane as components of receptors for viruses, peptide hormones, and pharmacologically active substances. They are composed of sphingosine, a long-chain fatty acid, hexose(s), and neuraminic acid or its *N*-acetyl derivative, sialic acid. Gangliosides do not contain glycerol, which forms the backbone of glycerides and phosphatides.

251. The answer is A. *(Mahler, ed 2. p 591.)* Chylomicrons, which have a triglyceride content in the range of 79 to 95 percent, have the lowest density among the substances listed in the question. Pre-β-lipoproteins contain 50 to 80 percent triglyceride and are known as very low density lipoproteins. β-Lipoproteins, which contain approximately 10% triglyceride, are known as low density lipoproteins. α-Lipoproteins contain 5 to 8 percent triglyceride and are referred to as high density lipoproteins. Transferrin, the iron-transport protein in plasma, is not a lipoprotein.

252. The answer is D. *(White, ed 6. pp 397-399.)* Cyanides inhibit the action of cytochrome oxidase, a key enzyme in the process of tissue respiration. In so doing, they reduce respiration by 60 to 90 percent, which indicates that this proportion of animal cell respiration involves the electron transport system terminating in cytochrome oxidase.

253. The answer is B. *(White, ed 6. pp 426-429.)* Glucose (and galactose) are transported against a concentration gradient by a carrier-mediated system exhibiting saturation kinetics, a system that functions only if there is concurrent movement of Na^+ from the intestinal lumen into the intestinal epithelial cell. The "uphill" transport of glucose seems to derive its energy not from metabolic sources but from the coupled movement of Na^+ "downhill," a gradient established by ATPase ejection of Na^+ from cells.

254. The answer is B. *(White, ed 6. pp 581-583.)* Phytanic acid is present in animal fat, cow's milk, and in foods derived from milk. The presence of the methyl group of carbon-3 blocks β-oxidation of this fatty acid. The normal oxidation of phytanic acid is by α-oxidation at carbon-2 followed by oxidative decarboxylation. The product, with a methyl group at carbon-2, can then undergo β-oxidation. In patients with Refsum's disease, the phytanic acid α-oxidase is nonfunctional.

255. The answer is C. *(McGilvery, pp 260, 296-299.)* Phosphorolysis of glycogen yields glucose 6-phosphate via glucose 1-phosphate. One molecule of ATP is utilized in the synthesis of fructose diphosphate. Two molecules of ATP are formed at each of two steps: phosphoglycerate kinase, and pyruvate kinase. Thus, there is a net yield of three ATP molecules per glycosyl residue converted to lactate.

256. The answer is D. *(Stryer, p 344.)* Glycolysis produces a net gain of 2 moles of ATP per mole of glucose utilized under anaerobic conditions. Under aerobic conditions of glycolysis, pyruvate dehydrogenase and the tricarboxylic acid cycle oxidations—coupled to the electron transport chain oxidative phosphorylations—produce a net gain of 32 moles of ATP per mole of glucose oxidized to CO_2 and H_2O. In addition, reducing equivalents (NADH) from glycolysis are oxidized by the glycerol phosphate shuttle which, because FAD is the mitochondrial electron carrier in this system, produces a net gain of 4 moles of ATP per mole of glucose utilized. Thus, under aerobic conditions there is a net gain of 36 moles of ATP per mole of glucose oxidized to CO_2 and H_2O.

257. The answer is D. *(Lehninger, ed 2. pp 536-538.)* A facultative cell, which can metabolize glucose under anaerobic or aerobic conditions, that is switched from anaerobic to aerobic metabolism experiences a sharp decrease in rate of glucose utilization. Under aerobic conditions, the cell can produce a net gain in moles of ATP formed per mole of glucose utilized that can be as high as 18 times that produced under anaerobic conditions. Thus the cell requires less glucose. Such increased ATP concentrations, together with the release of citrate from the tricarboxylic acid cycle under aerobic conditions, allosterically inhibit the key regulatory enzyme of the glycolytic pathway, phosphofructokinase.

258. The answer is D. *(Mahler, ed 2. p 512. White, ed 6. pp 216-218.)* Even in low concentration, fluoride is an inhibitor of enolase, an enzyme involved in the Embden-Meyerhof pathway of glycolysis. Fluoride combines with magnesium, calcium, and other divalent metals in the enzymes it affects, possibly by means of a fluorophosphate linkage.

259. The answer is B. *(McGilvery, p 260.)* Aerobic glycolysis can be defined as the oxidative conversion of glucose to two molecules of pyruvate. In the process, two molecules of ATP and two molecules of NADH are produced. Since reducing equivalents from the two molecules of NADH produced in the cytoplasm must be transported into the mitochondrion for oxidation, it is not known how many ATP are produced. On the assumption that two ATP molecules are formed per molecule of NADH oxidized via the glycerol phosphate shuttle, the ATP yield in aerobic glycolysis can be calculated as six ATP molecules per mole of glucose utilized.

260. The answer is D. *(Lehninger, ed 2. pp 381-382.)* Glycolysis takes place in the cytosol, which is the soluble region of the cytoplasm. The remaining energy-associated functions listed in the question occur in the mitochondria, which harbor the enzyme systems responsible for those functions.

261. The answer is C. *(Lehninger, ed 2. p 398.)* High-energy phosphates are defined in terms of an energy of hydrolysis greater than (more negative than) 7 kilocalories per mole. Except for glucose 6-phosphate ($\Delta G^{o'} = -3.3$ kilocalories per mole), all the compounds listed in the question have a free energy of hydrolysis more negative than -7.0 kilocalories per mole.

262. The answer is A. *(Lehninger, ed 2. pp 399-400.)* The terminal phosphate of ADP has a $\triangle G^{o\prime} = -7.3$ kilocalories per mole, about the same value as that for the hydrolysis of the terminal phosphate of ATP. What is of true significance is not the absolute value for $\triangle G^{o\prime}_{ATP}$ but its relation to the $\triangle G^{o\prime}$ of hydrolysis of phosphate group donors to ADP and to the $\triangle G^{o\prime}$ of phosphorylated compounds formed by phosphate transfer from ATP.

263. The answer is C. *(White, ed 6. pp 436-440.)* Fructose 1,6-diphosphate, one of the intermediaries in glycolysis, is synthesized from fructose 6-phosphate by phosphofructokinase; it constitutes "the committed step" in the glycolytic process. Aldolase catalyzes the reversible interconversion between fructose 1,6-diphosphate on the one hand, and glyceraldehyde 3-phosphate and dihydroxyacetone phosphate on the other.

264. The answer is B. *(Stryer, pp 310-311.)* A molecule of guanosine triphosphate is synthesized from guanosine diphosphate and phosphate at the cost of hydrolyzing succinyl CoA to succinate and CoA. This constitutes substrate-level synthesis of high-energy phosphate.

265. The answer is C. *(Lehninger, ed 2. pp 427-429.)* The two glycolytic reactions involving respectively an oxidation and a phosphate transfer, in which glyceraldehyde 3-phosphate is converted to 3-diphosphoglycerate, yield an overall energy change given by the expression:

$$\triangle G^{o\prime}\ total = \triangle G^{o\prime}{}_1 + \triangle G^{o\prime}{}_2 = +1.5 + (-4.5) = -3.0\ kcal/mol$$

266. The answer is A. *(Lehninger, ed 2. pp 436-437.)* Most of the fructose absorbed from the gut is phosphorylated by fructokinase at the C-1 position in the liver. The conversion of fructose 1-phosphate to fructose 6-phosphate does not occur, nor can phosphofructokinase utilize fructose 1-phosphate as a substrate. A specific aldolase (aldolase B) catalyzes the conversion of fructose 1-phosphate to dihydroxyacetone phosphate and glyceraldehyde.

267. The answer is A. *(Mahler, ed 2. pp 605-606.)* In the citric acid cycle, the step from citrate to isocitrate by way of *cis*-aconitate involves only the movement of a hydroxyl group and not donation of an H^+-e^- pair to the electron transport chain, as occurs in the other labeled steps on the diagram appearing in the question.

268. The answer is D. *(Stryer, p 315.)* Fumarase, aconitase, isocitrate dehydrogenase, and succinate thiokinase are all part of the citric acid cycle; pyruvate dehydrogenase is not. Pyruvate dehydrogenase is the enzyme complex that accomplishes oxidative decarboxylation of pyruvate to yield acetyl CoA.

269. The answer is D. *(Lehninger, ed 2. pp 366-367, 594-596.)* Fats, sugars, and ATP are the immediate sources of chemical energy for living cells; sunlight is an ultimate energy source (especially near the ultraviolet end of the visible spectrum), delivering photons utilized by chlorophyll and other plant pigments for photosynthesis. Heat, which can do work only when it is transferred from a warmer to a cooler body, is not available as an energy source for living organisms, which themselves are close to ambient temperature.

270. The answer is B. *(Lehninger, ed 2. pp 443-445.)* The intermediates α-ketoglutarate and oxaloacetate are formed by transamination of glutamic and aspartic acids. The citric acid cycle is an aerobic process providing the major catabolic pathway for glucose degradation and is a more efficient energy-producing process than glycolysis, which in normal cells is an anaerobic process.

271. The answer is C. *(White, ed 6. pp 338-341.)* The regulation of the citric acid cycle remains a matter of controversy; there is no universal agreement on which enzymes are involved in its regulation. Among the choices listed in the question, pyruvate dehydrogenase is not involved in the citric acid cycle and there is no enzyme called citrate dehydrogenase. Malate dehydrogenase and aconitase are thought to catalyze quasi-equilibrium reactions. Of the enzymes listed, isocitrate dehydrogenase is the only candidate for a role in regulation of the citric acid cycle.

272. The answer is E. *(White, ed 6. pp 426-427.)* In order of decreasing rate of absorption by the gastrointestinal tract, monosaccharides may be ranked: galactose > glucose > fructose > mannose > xylose > arabinose. Monosaccharides represent the final stage of carbohydrate digestion and are the product of disaccharidase action.

273. The answer is D. *(Mahler, ed 2. p 499.)* Glucokinase and hexokinase both catalyze the phosphorylation of glucose. While glucokinase is more specific for glucose than is hexokinase, the latter enzyme has a 1000-fold greater affinity for glucose. Only hexokinase is subject to feedback inhibition by glucose 6-phosphate. Glucokinase is present in liver but not in brain; hexokinase is present in both.

274. The answer is B. *(Mahler, ed 2. p 474.)* Cellulose is a high molecular weight, D-glucose polymer, which yields the disaccharide cellobiose (4-O-β-D-glucopyranosyl-D-α-glucopyranose) on partial hydrolysis. Lactose is a disaccharide of glucose and galactose, whereas maltose is a homopolysaccharide of glucose, and sucrose a disaccharide of fructose and glucose. Cellulose, a complex plant polysaccharide, is not digestible by humans.

275. The answer is D. *(McGilvery, pp 263-283.)* Lactate produced by glycolysis in peripheral tissues such as muscle, erythrocytes, and the renal medulla is taken up from the circulation by the liver and converted to glucose by the gluconeogenic pathway. This glucose is available for both maintenance of blood glucose levels and glycolysis in the peripheral tissues. The overall process requires an expenditure of four molecules of ATP per mole of glucose, thus completing the Cori cycle.

276. The answer is E. *(McGilvery, pp 322-331.)* In a process such as fatty acid synthesis, the reducing equivalents (NADPH) are thought to derive from two principal sources: the hexose monophosphate shunt, and the malate cycle (the transhydrogenation pathway involving malate dehydrogenase and malic enzyme).

277. The answer is E. *(McGilvery, p 280.)* For each molecule of lactate, one of ATP is utilized (either directly or indirectly) for each of the reactions catalyzed respectively by pyruvate carboxylase, phosphoenolpyruvate carboxykinase, and phosphoglycerate kinase. Thus, six molecules of ATP are required for the synthesis of one molecule of glucose from lactate.

278. The answer is D. *(Lehninger, ed 2. pp 624-627.)* All the compounds listed in the question except leucine yield a net synthesis of glucose through the tricarboxylic acid cycle and glycolysis; leucine is converted into CO_2 or acetyl CoA. It is, therefore, spoken of as a purely ketogenic amino acid, as opposed to the other compounds mentioned, which are glycogenic.

279. The answer is E. *(Mahler, ed 2. pp 384-385, 550.)* The following reaction,

$$UDP \text{ glucose} + \text{glycogen}_n \rightarrow UDP + \text{glycogen}_{n+1},$$

which is reversible, is catalyzed by UDP-sugar phosphorylases (uridyltransferases). The nucleotide uridine diphosphate, in this and other ways, is closely related to carbohydrate metabolism.

280. The answer is A. *(Mahler, ed 2. pp 499-500.)* In the following reaction,

ATP + glucose → glucose 6-phosphate + ADP,

ATP is the high-energy phosphate employed for phosphorylation of glucose via hexokinase or glucokinase. It is used also to phosphorylate fructose and fructose 6-phosphate.

281. The answer is C. *(Mahler, ed 2. p 520.)* In the reaction that follows,

oxaloacetate + GTP → GDP + phosphoenolpyruvate + CO_2,

GTP facilitates the decarboxylation of oxaloacetate and its entrance into a pathway essentially the reverse of glycolysis.

282. The answer is D. *(Mahler, ed 2. p 478.)* Hyaluronic acid is a heteropolysaccharide made up of D-glucuronic acid and N-acetyl-D-glycosamine units that alternate in position in the polysaccharide chain. The glucosamine-to-glucuronic acid link is β-(1 → 4) and the glucuronic acid-to-glucosamine link β-(1 → 3). Keratin is a protein. Chitin is composed of repeating N-acetyl-D-glucosamine residues, and both chondroitin sulfate and heparin contain sulfate groups.

283. The answer is E. *(Lehninger, ed 2. p 273.)* Hyaluronic acid, a linear polymer, is present in the connective tissue ground substance of vertebrates, in synovial fluid, and in vitreous humor. Two other acid mucopolysaccharides, chondroitin and keratan sulfate, occur in the cornea.

284. The answer is B. *(McGilvery, pp 579-582.)* The chondroitin chain is composed of alternating D-glucuronate and N-acetyl galactosamine residues. It is a precursor in the synthesis of protein polysaccharides and the final step in this synthesis is the addition of sulfate groups to the galactosamine residues. The sulfate donor is 3'-phosphoadenosine 5'-phosphosulfate, which is formed from ATP and inorganic sulfate by the action of adenosine phosphosulfate pyrophosphorylase. The reaction is driven by the rapid hydrolysis of the pyrophosphate group replaced by sulfate, and by phosphorylation of the ribose at the 3'-position by adenosine phosphosulfate kinase. Adenosine 5'-phosphosulfate will not donate its sulfate group to chondroitin, and sulfonyl urea does not have a sulfate group to transfer. Substituted sulfonyl ureas have been used as oral therapeutic agents in maturity-onset diabetes mellitus.

285. The answer is C. *(White, ed 6. p 46.)* Saponification is the name given to the alkaline hydrolysis of a neutral fat to yield a soap (salt of a fatty acid) and glycerol. It provides an initial analytic approach to neutral fats.

286. The answer is C. *(Lehninger, ed 2. pp 659-663.)* The addition of CO_2 to acetyl CoA yields malonyl CoA, from which fatty acids are synthesized in two carbon increments to chain lengths of 18 to 20 carbon atoms. This occurs outside of mitochondria, using NADP rather than NAD.

287. The answer is E. *(White, ed 6. p 602.)* A major fate of acetoacetyl CoA in the liver is the formation of 3-hydroxy-3-methylglutaryl CoA (HMG CoA) which is then cleaved to acetoacetate and acetyl CoA. In the cytoplasm, HMG CoA is a precursor for the synthesis of cholesterol.

288. The answer is C. *(White, ed 6. p 584-592.)* In the synthesis of fatty acids, acetyl CoA is synthesized within mitochondria and transported to the cytoplasmic compartment via citrate. Acetyl CoA carboxylase is thought to be the rate-limiting enzyme in the extramitochondrial portion of the lipogenic pathway. However, there is considerable evidence to suggest that the synthesis of acetyl CoA may be the rate-limiting step in the overall pathway of lipogenesis.

289. The answer is E. *(White, ed 6. pp 577-578.)* Thiokinase is involved in activation of fatty acids to CoA thioester derivatives. Acyl CoA dehydrogenase, enoyl hydrase, β-hydroxyacyl CoA dehydrogenase, and β-ketothiolase are all involved in the β-oxidation sequence, per se, in which the fatty acid chains are shortened by two carbon atoms.

290. The answer is D. *(McGilvery, p 321.)* Acetyl CoA within the mitochondrion is condensed with oxaloacetate to form citrate. Citrate is then transported out of the mitochondrion. In the cytoplasm, citrate is cleaved to acetyl CoA and oxaloacetate at the cost of hydrolyzing ATP.

291. The answer is A. *(White, ed 6. pp 577-579.)* Fatty acid oxidation occurs in mitochondria by successive two-carbon unit shortening of the aliphatic chain to produce ATP. This process is stimulated, not inhibited, by carnitine. The reactions are initiated by formation of the fatty acid thioester of CoA.

292. The answer is D. *(White, ed 6. pp 595-597.)* Two acyl CoAs (CoA derivatives of fatty acids) and glycerol 3-phosphate (derived by reduction of dihydroxyacetone phosphate or from phosphorylation of glycerol by glycerokinase and ATP) react to form phosphatidic acid. A phosphatase cleaves off the phosphate to give 1,2-diacylglycerol which then reacts with another acyl CoA to yield a triglyceride.

293. The answer is D (4). *(Mahler, ed 2. pp 698-699. White, ed 6. pp 361-365.)* Oligomycin prevents oxidative phosphorylation, inhibiting the phosphorylation of ADP to form ATP by mitochondrial ATPase. It prevents utilization of energy derived from electron transport for the synthesis of ATP. Oligomycin has no effect on coupling, but blocks mitochondrial phosphorylation so that both oxidation and phosphorylation cease in its presence.

294. The answer is B (1, 3). *(Lehninger, ed 2. p 258.)* In osazone formation, two molecules of phenylhydrazine react with the carbonyl function and the adjacent hydroxyl function in a sugar. In the cases of the commonly occurring hexoses, in which osazone formation abolishes configuration differences about carbon atoms 1 and 2 (e.g., glucose, mannose, and fructose), all give rise to the same osazones.

295. The answer is C (2, 4). *(Mahler, ed 2. pp 466-472. White, ed 6. pp 504-505.)* Ribulose and fructose are ketose sugars. Ribose and glucose are aldoses. Isomerases allow the interconversion of aldose and ketose sugars, such as glucose and fructose, via phosphoglucose isomerase.

296. The answer is E (all). *(White, ed 6. pp 169-171, 188-189, 324, 328.)* Nucleosides contain a base linked to D-ribose by glucosidic linkage at the C-1 carbon of the pentose. Nucleotides are phosphate esters of nucleosides. ATP is a nucleoside triphosphate, NAD a dinucleotide containing two bases (adenine and nicotinamide), and RNA a polynucleotide. Thus, all these compounds contain sugar (pentose) residues. In addition, acetyl coenzyme A (acetyl CoA) contains an adenosine 3'-phosphate 5'-pyrophosphate moiety and therefore also contains a sugar residue.

297. The answer is C (2, 4). *(McGilvery, pp 272-280, 344, 356-357, 369-373, 376-381.)* Serine and glycerol are metabolized to pyruvate and dihydroxy-acetone phosphate, respectively, both of which can be converted to glucose in mammalian liver. Neither oleic acid, which can be converted only to acetyl CoA but not to 3-carbon precursors of glucose, nor leucine is a substrate for gluconeogenesis.

298. The answer is D (4). *(White, ed 6. pp 33, 35.)* Glycogen is a highly branched, energy-storage molecule that is hydrolyzed by α-1,4-glucan 4-glucanhydrolase and α-1,4-glucan maltohydradase to yield, respectively, glucose and maltose (a disaccharide of glucose). Cellulose is a structurally important polymer of *O*-methylglucose.

299. The answer is A (1, 2, 3). *(Stryer, pp 378-380.)* Glycogen is a highly branched polymer of α-D-glucose residues linked in α-1,4-glycosidic linkage. The branch chains occur about every ten residues and are linked in α-1,6-glycosidic linkages.

300. The answer is A (1, 2, 3). *(White, ed 6. pp 462-463.)* In gluconeogenesis, the reactions catalyzed by hexokinase and phosphofructokinase are effectively reversed by the enzymes glucose 6-phosphatase and fructose 1,6-diphosphatase, respectively. The pyruvate kinase reaction is reversed by the joint actions of pyruvate carboxylase and phosphoenolpyruvate carboxykinase.

301. The answer is B (1, 3). *(Mahler, ed 2. pp 549-550.)* Glycogen synthetase is an enzyme which transfers glucosyl moieties from UDP-glucose to a glycogen polymer primer. The enzyme exists in two forms: an active, dephosphorylated form; and an inactive, phosphorylated form. In plants, ADP-glucose plays a role similar to that of UDP-glucose in animals.

302. The answer is D (4). *(White, ed 6. pp 466-467.)* Glucose 6-phosphate dehydrogenase is the first enzyme specifically involved in the pentose cycle (hexose monophosphate shunt). Fumarase and α-ketoglutarate dehydrogenase are citric acid cycle enzymes. Hexokinase is the initiating enzyme responsible for phosphorylation of glucose in the Embden-Meyerhof glycolysis pathway.

303. The answer is D (4). *(Stryer, pp 307-308.)* The conversion of pyruvate to acetyl CoA at the pyruvate dehydrogenase step means that the carbon from glucose is lost insofar as gluconeogenesis is concerned. The input of acetate carbons in the form of acetyl CoA into the citric acid cycle does not result in the formation of a net synthesis of citric acid cycle intermediates. Pyruvate carboxylase functions in gluconeogenesis with pyruvate as substrate, or as an anaplerotic reaction for the citric acid cycle. Glucose 6-phosphate dehydrogenase is the first step in the hexose monophosphate pathway leading to pentose formation and NADPH for reductive syntheses. Glyceraldehyde 3-phosphate dehydrogenase catalyzes a reversible step in the glycolytic pathway.

304. The answer is B (1, 3). *(Stryer, pp 316-320.)* NAD and FAD are two of the cofactors involved in the oxidative decarboxylation of pyruvate by pyruvate dehydrogenase. Another is thiamine pyrophosphate. Neither folic acid nor NADP is involved in pyruvate decarboxylation.

305. The answer is E (all). *(Lehninger, ed 2. pp 198, 452-453.)* Arsenite is an inhibitor of lipoic acid-containing enzymes like α-ketoglutarate dehydrogenase. Malonate is an inhibitor of succinate dehydrogenase, and fluoroacetate can be converted to fluorocitrate, which is an inhibitor of aconitase. The citric acid cycle requires oxygen and cannot proceed anaerobically.

306. The answer is A (1, 2, 3). *(Mahler, ed 2. pp 626-629.)* The regulation of the citric acid cycle is only incompletely understood. However, some factors of regulatory importance have been implicated: The level of oxaloacetate may affect the citrate synthetase reaction directly; NAD^+ is a substrate for isocitrate dehydrogenase, 2-oxoglutarate dehydrogenase, and malate dehydrogenase; the ADP/ATP ratio may affect the citrate synthetase reaction, as well as the reaction catalyzed by succinate thiokinase.

307. The answer is B (1, 3). *(White, ed 6. pp 333-334.)* The α-ketoglutarate dehydrogenase complex employs the cofactors NAD^+, FAD, thiamine pyrophosphate, lipoic acid, and CoA. This enzyme yields succinyl-CoA in a reaction that is virtually unidirectional because of energetics favoring the product.

308. The answer is B (1, 3). *(Mahler, ed 2. pp 495-498.)* Glycolysis can proceed anaerobically with the production of lactate from glucose. The NADH produced at the triose phosphate dehydrogenase step is reoxidized by way of lactate dehydrogenase, resulting in a regeneration of NAD^+.

309. The answer is B (1, 3). *(White, ed 6. pp 437-440.)* Citrate and ATP are two allosteric inhibitors of phosphofructokinase. The two other substances listed in the question, cyclic AMP and NH_4^+, both have been reported to increase the activity of phosphofructokinase under certain conditions in vitro.

310. The answer is A (1, 2, 3). *(Mahler, ed 2. pp 692-694.)* According to the Mitchell hypothesis of oxidative phosphorylation, protons are actively transported out of mitochondria with the energy supplied by the electron transport chain. As protons re-enter mitochondria down an electrochemical gradient, they provide the energy for ATP synthesis. Uncouplers render membranes "leaky" to protons. It is thought that the intramitochondrial pH is higher than the extramitochondrial pH.

311. The answer is C (2, 4). *(Mahler, ed 2. pp 35-40.)* "Energy-rich" compounds are those that have a free energy of hydrolysis that is large in absolute value and negative in sign. In some instances (ATP), the availability of more resonance forms for the products of hydrolysis (ADP and inorganic phosphate) contributes to the highly negative free energy of hydrolysis. Because of the ionization of the product, esters of amino acids are high-energy compounds.

312. The answer is C (2, 4). *(Stryer, pp 341-343.)* The inner membrane of the mitochondrion is impermeable to NADH itself. Reducing equivalents can be transferred from the cytoplasm across the inner membrane by the glycerol 3-phosphate shuttle. D-Hydroxyacetone formed in glycolysis is reduced by NADH to glycerol 3-phosphate which can cross the inner membrane. In the mitochondrion, the glycerol 3-phosphate is reoxidized by an FAD-linked dehydrogenase to dihydroxyacetone which can return to the cytoplasm. Because the mitochondrial dehydrogenase is FAD-linked rather than NAD^+-linked, only two moles of ATP are formed per mole of reducing equivalent. It is thought that this shuttle operates more readily than a similar oxaloacetate-malate shuttle, in which both cytoplasmic and mitochondrial dehydrogenase are NAD^+-linked, because the high intramitochondrial levels of NADH favor the FAD-linked shuttle.

313. The answer is B (1, 3). *(Stryer, pp 390-391.)* Glycogen phosphorylase is subject to a variety of modes of regulation. The enzyme exists in two forms: phosphorylase *b*, which is inactive and unphosphorylated, and phosphorylase *a*, which is active and phosphorylated. Phosphorylase *b* is an allosteric enzyme and exhibits activity in the presence of its activator, AMP. Phosphorylase *b* is phosphorylated to phosphorylase *a* by phosphorylase *b* kinase, an enzyme that is itself activated by a cyclic AMP-dependent protein kinase. Phosphorylase *b* kinase also is activated by Ca^{++}, a regulation important in skeletal muscle inasmuch as muscle contraction is triggered by the release of Ca^{++}

314. The answer is E (all). *(Mahler, ed 2. p 24.)* Hydrolysis of the phosphate in all the compounds listed in the question, including ADP $\rightarrow$ AMP + P_i, yields a free-energy change of more than -7 kilocalories/mole. Free-energy changes exceeding -5 kilocalories/mole in hydrolysis of phosphate bonds are considered to be "high."

315. The answer is C (2, 4). *(Mahler, ed 2. pp 473-479.)* Hyaluronic acid and heparin are both mucopolysaccharides; cellulose is a polysaccharide. Neuraminidase is an enzyme composed not of mucopolysaccharide but of protein.

316. The answer is C (2, 4). *(White, ed 6. pp 351-357.)* Oxidative phosphorylation can be uncoupled from respiration by chemicals like dinitrophenol without affecting the function of electron transport. $\Delta G^{o\prime} = -52.7$ kilocalories for the transfer of a pair of electrons from NADH to O_2.

317. The answer is A (1, 2, 3). *(Mahler, ed 2. pp 473-479.)* Polysaccharides, which are used as energy storage forms (starch, glycogen), exist as both linear (cellulose) and branched (glycogen) polymers. They form components of bacterial cell walls but, because most polysaccharides have only one or two repeating units, unlike proteins and nucleic acids they cannot function as informational molecules.

318. The answer is B (1, 3). *(White, ed 6. p 61.)* Bile acids often are conjugated with glycine to form glycocholic acid, and with taurine to form taurocholic acid. In human bile, glycocholic acid is by far the more common.

319. The answer is A (1, 2, 3). *(White, ed 6. pp 38-69.)* Lipids are hydrophobic, water-insoluble substances that may contain phosphate and nitrogen in addition to carbon, hydrogen, and oxygen. They exhibit a multifarious structure, ranging from fatty acids to lipopolysaccharides.

320. The answer is C (2, 4). *(Stryer, p 291.)* ATP is synthesized by two reactions in glycolysis: the reactions catalyzed by phosphoglycerate kinase (reaction F to 2-phosphoglycerate in the pathway exhibited in the question) and pyruvate kinase (reaction G-H). Compound A is glucose 6-phosphate; B is fructose 6-phosphate; C is fructose 1,6-diphosphate; D is glyceraldehyde 3-phosphate; E is 1,3-diphosphoglycerate; F is 3-phosphoglycerate; G is phosphoenol pyruvate; and H is pyruvate.

321. The answer is A (1, 2, 3). *(White, ed 6. pp 602-603, 1292.)* Starvation results in the increased use of lipids as an energy source, with increased fatty acid oxidation and production of acetoacetyl CoA, a precursor of ketone bodies. The utilization of ketone bodies for metabolism is also seen in insulin deficiency (diabetes mellitus), but not in diabetes insipidus. Diabetes insipidus results from a lack of vasopressin, and is manifested by polyuria and polydipsia.

322. The answer is C (2, 4). *(Stryer, pp 418-419, 426.)* Pyruvate and acetyl CoA are important intermediates in the conversion of glucose to fatty acids. The acetyl CoA exits from the mitochondrion in the form of citrate, not as acetyl carnitine. Acetyl carnitine occurs in the mitochondrial fatty acid oxidative process.

323. The answer is D (4). *(White, ed 6. pp 593-594.)* Linoleic and linolenic fatty acids cannot be synthesized by mammals and are, therefore, referred to as essential. Other fatty acids, including palmitic, stearic, and oleic acids, that do not contain double bonds between the seventh carbon from the terminal methyl group and the carboxyl group, **can** be made by alternate desaturation and elongation in mammals.

324. The answer is B (1, 3). *(Mahler, ed 2. pp 592-599, 714-722.)* In lipogenesis, NADPH is the source of reducing equivalents: in fatty acid oxidation, both FAD and NAD^+ serve as cofactors. In lipogenesis, both the acyl carrier protein and CoA are pantetheine-containing coenzymes, or prosthetic groups, involved in the pathway. In lipogenesis, acetyl CoA is carboxylated to yield malonyl CoA in order to activate the acetate moiety for addition to the growing fatty acid. The reactions of lipogenesis proceed **extra**mitochondrially: fatty acids are oxidized **within** the mitochondria.

325. The answer is E (all). *(White, ed 6. pp 583-584.)* Propionic acid is activated to propionyl CoA which is carboxylated to D-methylmalonyl CoA. After racemization to L-methylmalonyl CoA, the molecule is rearranged by way of a vitamin B_{12}-dependent reaction to give succinyl CoA.

326. The answer is C (2, 4). *(Lindros, J Biol Chem 249 [1974]:7956-7963.)* The principal pathway for hepatic ethanol metabolism is thought to be oxidation to acetaldehyde in the cytoplasm by alcohol dehydrogenase. Acetaldehyde is then oxidized, probably within the mitochondrion, to yield acetyl CoA. Neither acetone nor methanol appears in this biodegradation pathway.

327. The answer is A (1, 2, 3). *(White, ed 6. pp 619-626.)* Cholesterol is the precursor of progesterone in a series of reactions leading from cholesterol to progesterone to the sex hormones and adrenal corticosteroids. Mevalonic acid is an early precursor, and squalene, lanosterol, zymosterol, and desmosterol, are all late precursors of cholesterol.

328. The answer is C (2, 4). *(White, ed 6. pp 1221-1229.)* Cortisol and aldosterone have 11β-hydroxyl functions. Estradiol and testosterone lack an 11-OH function, but possess other hydroxyl functions which are individually specific.

329. The answer is A (1, 2, 3). *(Harper, ed 16. pp 486-488.)* Hydroxylation at the 17-position is a part of the biosynthetic pathways leading to estradiol, cortisol, and testosterone. Aldosterone is hydroxylated at the 18-position, and then desaturated to form an aldehyde. All steroid biosynthesis involves successive hydroxylation of cholesterol and its derivatives, commencing with pregnenolone.

330. The answer is C (2, 4). *(Stryer, pp 497-498.)* In general, C-21 hydroxylation is necessary for both glucocorticoid and mineralocorticoid activity. There is a 21-hydroxylase that is required to hydroxylate at C-21 in the biosynthesis of cortisol and aldosterone. Androgens, in general, lack carbons 20, 21 of the steroid nucleus, and so are designated C-19 steroids. Estrogens are C-18 steroids.

331. The answer is E (all). *(White, ed 6. pp 48-52, 607-611.)* Phosphoglycerides contain glycerol phosphate, two fatty acid molecules, and either choline, ethanolamine, serine, or inositol bound in an ester linkage to phosphoric acid. Depending upon their constitution, they may be designated specifically phosphatidyl choline, phosphatidyl ethanolamine, phosphatidyl serine, and phosphoinositide.

332. The answer is E (all). *(White, ed 6. pp 572-573.)* Chylomicrons contain triglycerides (79-95 percent), cholesterol (1-5 percent), phospholipids (3-15 percent), and protein (0.5-2.5 percent). They are found in lymphatic chyle during absorption of fat from the gut and are discharged into the venous blood via the thoracic duct.

333. The answer is D (4). *(Jawetz, ed 13. p 61.)* The glyoxylate cycle appears in microorganisms that are furnished only acetate as a source of carbon for growth, under aerobic conditions. Acetate is metabolized to succinate, and so provides the source of C_4 compounds that cannot, under these conditions, be provided by the citric acid cycle.

334. The answer is C (2, 4). *(White, ed 6. pp 532-534.)* Biosynthesis of starch during photosynthesis occurs via the Calvin cycle (pentose phosphate pathway). CO_2 fixation in this reaction pathway has been demonstrated to occur in the dark, after light-priming of chloroplasts, yielding 3-phosphoglyceric acid from ribulose diphosphate and CO_2.

335. The answer is A (1, 2, 3). *(White, ed 6. pp 489, 541.)* Levan is produced in *Bacillus megatherium* by levansucrase, a transglycosylase, from sucrose. It is a fructose polymer similar to dextran, a glucose polymer. Formation of both polymers exhibits strong similarities to the action of the branching enzyme amylotransglycosylase in the synthesis of glycogen.

336. The answer is C (2, 4). *(White, ed 6. p 541.)* Bacterial transglycosylases can convert disaccharides to linear polymers of one of the disaccharide units. For example, *Leuconostoc mesenteroides* converts sucrose to dextran and fructose. Dextran is a glucose polymer in which there is a glycosidic linkage between carbon 1 of each glucose unit and carbon 6 of the next residue. Another bacterial enzyme converts sucrose to levan and glucose, levan being a linear polymer of fructose in 2,6-fructosidic linkage. Dextran and levan are present in dental plaque, a condition favored by a high dietary intake of sucrose.

337. The answer is C (2, 4). *(Davis, ed 2. p 117.)* Lipid A is the base on which the core polysaccharide forms. Ketodeoxyoctonic acid is considered to be part of the lipid A complex. The core polysaccharides are similar for members of each group of bacteria in the Enterobacteriaceae, and can be used to define them antigenically.

338. The answer is D (4). *(White, ed 6. pp 577-584.)* Beta-oxidation of fatty acids (the major fate of fatty acids) proceeds within mitochondria. Fatty acid oxidation by other routes, i.e., α- and ω-oxidation, may proceed in the microsomal fraction of cells.

339. The answer is D (4). *(Stryer, pp 316-320.)* Lipoic acid is an essential cofactor for enzyme complexes involved in the oxidative decarboxylation of α-keto acids, e.g., pyruvate, α-ketoglutarate, and the α-keto acids formed by transamination of the branched chain amino acids (leucine, isoleucine, valine). The enzymes of the glycolytic pathway do not require lipoic acid as a cofactor. Transaminases require pyridoxal phosphate and carboxylases require biotin.

340. The answer is C (2, 4). *(Lehninger, ed 2. pp 424-425, 430-431, 661-663.)* Citrate activates acetyl CoA carboxylase but inhibits phosphofructokinase and pyruvate kinases (in liver). Citrate does not have a regulatory role in the enolase catalyzed reaction. Phosphofructokinase is the rate-limiting step in the glycolytic pathway, while acetyl CoA carboxylase catalyzes the rate-limiting step in fatty acid synthesis.

341. The answer is D (4). *(White, ed 6. pp 593-595.)* The mammalian poly-desaturase enzyme is extramitochondrial and membrane-bound (sedimentable at 100,000 g for 60 minutes). Microsomal enzyme systems also exist for elongation of saturated and unsaturated fatty acyl CoA derivatives.

342. The answer is B (1, 3). *(Williams, ed 5. pp 248-249.)* 11-Ketoetiocholanolone and androsterone both are 17-ketosteroids. In the former, the presence of an 11-keto function suggests an adrenal rather than gonadal origin because the adrenal gland has an 11-hydroxylase enzyme that is absent from ovary and testis.

343. The answer is D (4). *(Williams, ed 5. pp 248-249.)* The Porter-Silber reaction requires hydroxyl functions at both C-17 and C-21, as well as a 20-keto function. Therefore, while tetrahydrocortisol is a "17-hydroxysteroid," cortol, despite the presence of a 17-hydroxyl function, lacks a keto function and so is not.

344. The answer is C (2, 4). *(Williams, ed 5. pp 249-250.)* 17-Ketogenic steroids are those that can be converted to 17-ketosteroids by oxidation with sodium bismuthate. This reaction requires a hydroxyl group at the 17-position as well as at either C-20 or C-21. Thus, both cortol and tetrahydrocortisol, two degradation products of cortisol, are 17-ketogenic steroids.

345-347. The answers are: 345-C, 346-D, 347-B. *(Mahler, ed 2. pp 472-474.)* Sucrose, the chief dietary sugar, is a disaccharide of glucose and fructose (α-D-glucopyranosyl-(1 → 2)-β-D-fructofuranoside). Lactose, or milk sugar, is a disaccharide of galactose and glucose (β-D-galactopyranosyl-(1→4)-β-D-glucopyranose). Maltose is a dissacharide that arises as a hydrolytic product of starch, and is composed of two glucose units (α-D-glucopyranosyl-(1 → 4)-β-D-glucopyranose).

348-349. The answers are: 348-E, 349-A. *(Mahler, ed 2. pp 475-484.)* Mureins are polysaccharides that constitute the principal components of cell walls of both gram-negative and gram-positive bacteria. Keratan sulfate is a principal polysaccharide component of connective tissue. Starches are a storage form of intracellular polysaccharides, in plants, while cellulose is a major structural polysaccharide in plants. Teichoic acids are polymers of either glycerol or ribotol residues linked by 1,3-phosphodiester bridges.

350-352. The answers are: 350-A, 351-E, 352-B. *(Stryer, p 291.)* ATP transfers its terminal phosphate to fructose 6-phosphate, producing fructose 1,6-diphosphate in a reaction which also requires Mg^{++}. Pyruvate is reduced to lactate by NADH in a reaction that has a large negative free-energy change. Glyceraldehyde 3-phosphate is oxidized by NAD to the corresponding acid which is phosphorylated with inorganic phosphate to yield 1,3-diphosphoglycerate.

353-355. The answers are: 353-A, 354-B, 355-C. *(Lehninger, ed 2. p 300. Williams, ed 5. pp 854-867.)* Structurally, the compounds exhibited in the question should easily be identified as prostaglandins. If the structure is not sufficient for recognition, the supplementary information that PGA does not stimulate nonvascular smooth muscle, and that PGF causes a transient increase in blood pressure, should make identification of these substances quite simple. Several prostaglandins have oxytocic application, including the induction of therapeutic abortion.

Vitamins and Hormones

DIRECTIONS: Each question below contains five suggested answers. Choose the **one best** response to each questions.

356. A pellagra-like skin rash may be seen in

(A) phenylketonuria
(B) homogentisic aciduria
(C) Hurler's syndrome
(D) Hartnup disease
(E) gargoylism

357. The most important signal for the release of insulin from pancreatic islet beta-cells is

(A) increased intestinal motility
(B) increased blood glucose level
(C) increased blood lipid level
(D) hypothalamic stimulation
(E) release of epinephrine from the adrenal medulla

358. The metal cofactor of tyrosinase is

(A) cobalt
(B) copper
(C) iron
(D) manganese
(E) magnesium

359. All the following hormones or compounds cause decreased lipolysis EXCEPT

(A) insulin
(B) prostaglandin E
(C) nicotinic acid
(D) beta-adrenergic blockers
(E) glucagon

360. In prolonged fasting, a limiting factor in the amount of glucose produced by the liver is

(A) guanidine
(B) alanine
(C) tryosine
(D) tryptophan
(E) cytosine

361. Pairs of hormones with antagonistic effects include all the following EXCEPT

(A) vasopressin — oxytocin
(B) insulin — glucagon
(C) melanocyte-stimulating hormone — melatonin
(D) calcitonin — parathyroid hormone
(E) histamine — serotonin

362. Insulin has many direct effects on various cell types from such tissues as muscle, fat, liver, and skin. All the following cellular activities are increased following exposure to physiologic concentrations of insulin EXCEPT

(A) plasma membrane transfer of glucose
(B) glucose oxidation
(C) gluconeogenesis
(D) lipogenesis
(E) formation of ATP, DNA, and RNA

363. All the following cellular activities are decreased following exposure to physiologic concentrations of insulin EXCEPT

(A) ureogenesis
(B) proteolysis
(C) keotgenesis
(D) glycogenolysis
(E) proteogenesis

364. Which of the hormones listed below is produced by cells in the pancreatic islets of Langerhans?

(A) Somatomedin
(B) Somatostatin
(C) Somatotropin
(D) Cholecystokinin
(E) Secretin

365. Oxytocin, originally assayed by its ability to cause uterine muscle to contract, literally means "to stimulate birth." It is now known that oxytocin plays only a secondary role in uterine labor and that its major role is to

(A) increase blood pressure
(B) decrease blood pressure
(C) cause the placenta to release from the uterine wall
(D) cause the release of milk in the lactating breast
(E) control capillary permeability to oxygen in the lungs

366. All the following statements about prostaglandins are true EXCEPT that

(A) they are cyclic fatty acids
(B) they have potent biologic effects that involve almost every organ in the body
(C) they were first observed to cause uterine muscle contraction and lowering of blood pressure
(D) although found in many organs, they are synthesized only in the prostate and seminal vesicles
(E) medullin, isolated from the kidney medulla, is a prostaglandin

367. In normal humans, <u>insulin secretion</u> during constant <u>intravenous glucose</u> administration is best characterized by which of the following curves? (Glucose administration starts at time = 0.)

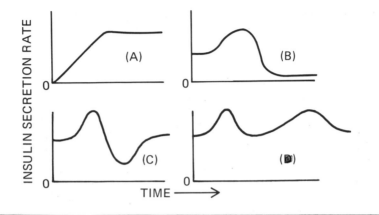

368. <u>Corneal vascularization</u> may be a sign of deficiency of

(A) thiamine
(B) riboflavin
(C) niacin
(D) vitamin A
(E) inositol

369. The hormones that <u>differ LEAST</u> in chemical structure are

(A) insulin and proinsulin
(B) oxytocin and vasopressin (ADH)
(C) cortisone and growth-stimulating hormone
(D) thyrotropin and oxytocin
(E) insulin and cortisone

370. A patient's glucose tolerance test is shown below. The most likely diagnosis is

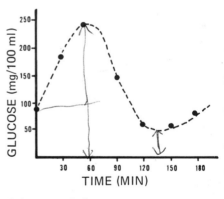

(A) severe diabetes
(B) Addison's disease
(C) von Gierke's disease
(D) delayed insulin response
(E) normal insulin response

Questions 371-372

371. The structure shown below is a

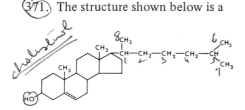

(A) hormone
(B) cerebroside
(C) steroid
(D) terpene
(E) vitamin

372. The central ring structure shown above is common to all the following compounds EXCEPT

(A) adrenocorticotropin
(B) aldosterone
(C) testosterone
(D) bile acids
(E) vitamin D

373. All the following hormones are elaborated by the anterior pituitary in mammals EXCEPT

(A) somatotropin
(B) vasopressin (A D H)
(C) luteinizing hormone
(D) follicle-stimulating hormone
(E) prolactin

374. Molecular iron, Fe, is

(A) stored primarily in the spleen
(B) excreted in the urine as Fe^{++}
(C) stored in the body in combination with ferritin
(D) absorbed in the intestine by transferrin
(E) absorbed in the ferric, Fe^{+3}, form

375. A lack of which of the following nutrients is best tolerated by humans?

(A) Protein
(B) Iodine
(C) Carbohydrate
(D) Lipid
(E) Calcium

376. All the following compounds are members of the electron transport chain EXCEPT

(A) ubiquinone (coenzyme Q)
(B) cytochrome c
(C) NAD
(D) FAD
(E) carnitine

377. A deficiency of vitamin B_{12} causes

(A) cheilosis
(B) beriberi
(C) pernicious anemia
(D) scurvy
(E) rickets

378. In adults, a severe deficiency of vitamin D causes

(A) night blindness
(B) osteomalacia
(C) rickets (vit D dy in children)
(D) skin cancer
(E) none of the above

379. Classical scurvy results from a deficiency of

(A) thiamine
(B) riboflavin
(C) pantothenic acid
(D) ascorbic acid
(E) vitamin A

380. The coenzyme required in oxidative decarboxylation is

(A) biotin
(B) 5′-deoxyadenosylcobalamin
(C) pyridoxal phosphate
(D) ascorbic acid
(E) thiamine pyrophosphate

381. Pellagra can be prevented by treatment with

(A) thiamine
(B) niacin
(C) pyridoxine
(D) vitamin B_{12}
(E) pantothenic acid

382. Blockage to absorption of fat-soluble vitamins could cause which of the following disorders?

(A) Pernicious anemia
(B) Scurvy
(C) Pellagra
(D) Rickets
(E) Beriberi

383. Both Wernicke's disease and beriberi can be reversed by administering

(A) vitamin C
(B) vitamin A
(C) thiamine (B_1)
(D) vitamin B_6
(E) vitamin B_{12}

384. A coenzyme that functions with thiamine pyrophosphate in the decarboxylation of α-keto acids to yield acyl CoA compounds is

(A) biotin
(B) lipoic acid
(C) vitamin A
(D) vitamin C
(E) NADP

385. The action of which of the following water-soluble vitamins is antagonized by methotrexate?

(A) Vitamin B_{12}
(B) Riboflavin
(C) Vitamin C
(D) Vitamin B_6 (pyridoxine)
(E) Folic acid

386. Pantothenic acid is a constituent of the coenzyme involved in

(A) decarboxylation
(B) acetylation
(C) dehydrogenation
(D) reduction
(E) oxidation

387. Which of the following polypeptide hormones shows the LEAST
structural similarity to all the others
listed?

(A) Follicle-stimulating hormone
(FSH)
(B) Adrenocorticotrophic hormone
(ACTH)
(C) Luteinizing hormone (LH)
(D) Human chorionic gonadotropin
(HCG)
(E) Thyroid-stimulating hormone
(TSH)

388. All the following cofactors are
involved in the pyruvate dehydrogenase
reaction EXCEPT

(A) pyridoxal phosphate
(B) thiamine pyrophosphate
(C) lipoic acid
(D) FAD
(E) CoA

389. The administration of raw egg
white to animals induces a deficiency
of which of the following vitamins?

(A) Pantothenate
(B) Riboflavin
(C) Thiamine
(D) Ascorbate
(E) Biotin

390. Biotin is involved in which of
the following types of reactions?

(A) Hydroxylations
(B) Carboxylations
(C) Decarboxylations
(D) Dehydrations
(E) Deaminations

391. The administration of avidin, a
biotin antagonist, would be expected
to affect reactions catalyzed by all
the following enzymes EXCEPT

(A) succinic thiokinase _Not carboxylase_
(B) propionyl CoA carboxylase
(C) β-methylcrotonyl CoA car-
boxylase
(D) acetyl CoA carboxylase
(E) pyruvic acid carboxylase

392. Which of the following vitamins
is required for the action of transaminases?

(A) Niacin
(B) Pantothenate
(C) Thiamine
(D) Pyridoxal phosphate
(E) Riboflavin

393. Which of the following compounds is synthesized from glutamic
acid, para-aminobenzoic acid, and a
pteridine nucleus?

(A) Vitamin B_{12}
(B) Cyanocobalamin
(C) Folic acid
(D) Biotin
(E) CoA

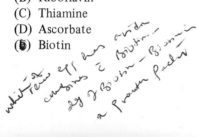

Questions 394-395

394. Which of the structures shown below is specific for the treatment of acute pellagra?

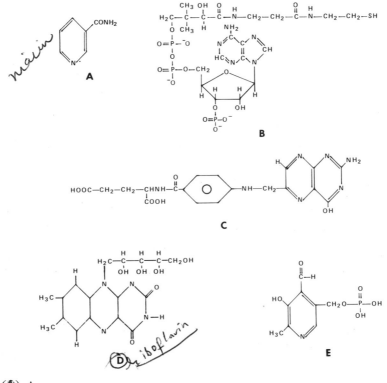

(A) A
(B) B
(C) C
(D) D
(E) E

395. A deficiency of compound D in the figure above can cause

(A) convulsive seizures
(B) liver damage
(C) pernicious anemia
(D) beriberi
(E) cheilosis

396. Which of the following vitamins is the precursor of coenzyme A?

(A) Riboflavin
(B) Pantothenate
(C) Thiamine
(D) Cobamide
(E) Pyridoxamine

397. Vitamin E is a

(A) fatty acid
(B) propylthiouracil analog
(C) tocopherol
(D) quinone
(E) prostaglandin

DIRECTIONS: Each question below contains four suggested answers of which **one** or **more** is correct. Choose the answer:

A	if	**1, 2, and 3**	are correct
B	if	**1 and 3**	are correct
C	if	**2 and 4**	are correct
D	if	**4**	is correct
E	if	**1, 2, 3, and 4**	are correct

398. The effects of glucagon include the

(1) activation of phosphorylase kinase
(2) inhibition of gluconeogenesis
(3) production of glycogenolysis
(4) activation of muscle phosphorylase

399. Which of the following hormones are derived from the same amino acid?

(1) Epinephrine
(2) Vasopressin
(3) Thyroxine
(4) Prostaglandin E

400. Human chorionic gonadotropin (HCG) has one peptide chain in common with which of the following hormones?

(1) LH
(2) TSH
(3) FSH
(4) PGH (human placental lactogen)

401. Which of the following statements about the insulin molecule are true?

(1) After denaturation, it does not spontaneously renature to produce fully active molecules in good yield
(2) It is assembled from two fragmentary polypeptide chains
(3) It consists of two different polypeptide chains linked together by disulfide bonds
(4) Its subunit polypeptide chains are each biologically active with one-half the potency of the intact hormone

402. Coenzymes derived from the vitamin shown below are required by which of the following enzymes?

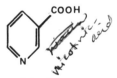

(1) Lactate dehydrogenase
(2) UDP glucose epimerase
(3) Phenylalanine hydroxylase
(4) Polynucleotide ligase of *E. coli*

403. Coenzymes derived from the vitamin shown below are required by which of the following enzymes?

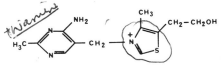

(1) Pyruvate dehydrogenase
(2) Transketolase (of pentose cycle)
(3) α-Ketoglutarate dehydrogenase
(4) Glutamate dehydrogenase

404. Coenzymes derived from the vitamin shown below are required by which of the following enzymes?

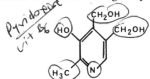

(1) Glutamic-oxaloacetic transaminase
(2) Glycogen phosphorylase
(3) Dihydroxyphenylalanine decarboxylase
(4) Aldolase

405. Coenzymes derived from the vitamin shown below are required by which of the following enzymes?

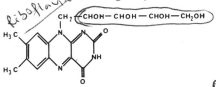

(1) Glutamate dehydrogenase
(2) Pyruvate dehydrogenase
(3) Lactate dehydrogenase
(4) Succinate dehydrogenase

406. Which of the following erythrocyte enzymes can serve as a sensitive indicator of thiamine deficiency?

(1) Glucose 6-phosphate dehydrogenase
(2) Transaldolase
(3) 2,3-Diphosphoglycerate mutase
(4) Transketolase

407. Fat-soluble vitamins include

(1) retinol (vit A)
(2) calciferol vit D
(3) tocopherol vit E
(4) ascorbic acid

408. Vitamin B_6 (pyridoxine) is frequently a precursor of a cofactor in

(1) decarboxylation
(2) deamination
(3) transamination
(4) transsulfuration

409. Hormones secreted by the pituitary include

(1) ACTH
(2) aldosterone
(3) FSH
(4) parathormone

410. Which of the following are water-soluble vitamins?

(1) Vitamin D
(2) Vitamin K
(3) Vitamin A
(4) Vitamin B_{12}

SUMMARY OF DIRECTIONS

A	B	C	D	E
1, 2, 3 only	1, 3 only	2, 4 only	4 only	All are correct

411. NAD, NADP, CoA, FAD, and ATP all contain

(1) at least one phosphate
(2) adenosine
(3) ribose
(4) a vitamin

412. Porphyrin-containing coenzymes or prosthetic groups are contained in which of the following proteins?

(1) Hemoglobin
(2) Methylmalonyl CoA mutase
(3) Cytochrome *c*
(4) Ferredoxin

414. Which of the following polypeptide hormones contain six amino acids in an S-S ring (with a disulfide bridge between the first and the last of the six amino acids)?

(1) Insulin
(2) Oxytocin
(3) Vasopressin
(4) Glucagon

415. Disclosure of the modes of action of hormones involves the use of antimetabolites or specific blocking agents to inhibit reactions at certain steps. If puromycin and cycloheximide blocked the action of a certain hormone, and actinomycin D and mitomycin C did not, it could be assumed that the mode of action of the hormone required

(1) active membrane transport of specific amino acids
(2) DNA synthesis
(3) RNA synthesis
(4) protein synthesis

413. Coenzymes derived from the vitamin shown below are required by enzymes involved in the *de novo* synthesis of which of the following nucleotides?

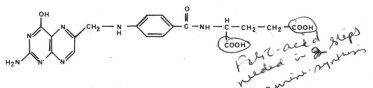

(1) Adenosine triphosphate (ATP)
(2) Guanosine triphosphate (GTP)
(3) Thymidine triphosphate (TTP)
(4) Cytidine triphosphate (CTP)

DIRECTIONS: The groups of questions below consist of lettered choices followed by several numbered items. For each numbered item select the **one** lettered choice with which it is **most** closely associated. Each lettered choice may be used once, more than once, or not at all.

Questions 416-418

For each hormone that follows, select the endocrine gland that produces it.

(A) Adenohypophysis
(B) Neurohypophysis
(C) Adrenal medulla
(D) Adrenal cortex
(E) Thyroid

A 416. ACTH

C 417. Norepinephrine

E 418. Calcitonin

Questions 419-421

For each external secretion that follows, select the protein that governs it.

(A) Cholecystokinin
(B) Gastrin
(C) Insulin
(D) Intrinsic factor
(E) Secretin

E 419. Secretion of pancreatic juice into the intestine

A 420. Release of bile from the gallbladder

B 421. Secretion of acid in the stomach

Questions 422-424

For each external secretion that follows, select the component with which it is most closely identified.

(A) Calcium ion
(B) Bicarbonate ion
(C) Hydrogen ion
(D) Epinephrine
(E) Lysozyme

B 422. Pancreatic secretion

C 423. Gastric juice

E 424. Tears

Questions 425-427

For each of the steroid-derived compounds that follow, choose the organ in which it is synthesized.

(A) Skin
(B) Bone
(C) Liver
(D) Kidney
(E) Small intestine

A 425. Cholecalciferol

C 426. 25-Hydroxycholecalciferol

D 427. 1,25-Dihydroxycholecalciferol

Vitamins and Hormones

Answers

356. The answer is D. *(Goodhart, ed 5. p 1034.)* Hartnup disease consists of a pellagra-like rash following exposure to sunlight, along with intermittent cerebellar ataxia, psychosis, and a massive aminoaciduria. Large amounts of indole 3-acetic acid and indican also are excreted. The biochemical defect involves a failure of the transport system handling monoamino monocarboxylic acids, although the transport of lysine and possible histidine also is affected. Affected patients have diminished renal tubular and intestinal (jejunum) absorption of these particular amino acids. These individuals also suffer a diminished ability to convert tryptophan to kynurenine and nicotinamide, a deficit that is responsible for the pellagra-like (niacin deficiency) symptoms.

357. The answer is B. *(Williams, ed 5. pp 512-516.)* An increase in level of blood glucose is the major signal for the release of insulin. Amino acids and, to some extent, free fatty acids also may trigger insulin release, but to a lesser degree than glucose. Insulin also is subject to release in the basal, as opposed to the exogenously stimulated, state.

358. The answer is B. *(White, ed 6. pp 408, 412, 743.)* Tyrosinase, which catalyzes the oxidation of phenol derivatives to quinones, contains the copper ion. Copper also is seen in electron transport enzymes, and in oxidases of, *inter alia*, ascorbate, laccase, and galactose.

359. The answer is E. *(Williams, ed 5. p 536.)* Glucagon is known to promote lipolysis and mobilize energy stored in fatty tissues. Insulin, beta-adrenergic blockers, and nicotinic acid all lower the blood level of fatty acids, and decrease lipolysis. Prostaglandin E, in low dosages, increases lipolysis, while in larger doses it decreases lipolysis.

360. The answer is B. *(Williams, ed 5. p 544.)* After two days of fasting, liver and muscle glycogen supplies essentially are gone and amino acids become the prime precursors for glucose formation (gluconeogenesis). Most of the amino acids must then be taken from protein that is serving an important role in cell structure, enzyme activities, and other functions. A significant amount of alanine no longer can be derived from pyruvate. After prolonged starvation, gluconeogenesis decreases and the plasma alanine level is less than after an overnight fast. That this is a limiting factor in the amount of glucose produced by the liver is substantiated by the fact that infusion of alanine promptly increases the blood glucose level.

361. The answer is A. *(White, ed 6. pp 720, 722, 1213-1218, 1274-1276, 1291-1294.)* Both vasopressin and oxytocin are secreted by the neurohypophysis. Vasopressin is a pressor and antidiuretic and oxytocin primarily stimulates lactation. The other pairs of hormones cited in the question are antagonists involved in the regulation of: B - glucose homeostasis; C - melanization; D - calcium metabolism, and E - vasomotor tone.

362. The answer is C. *(Williams, ed 5. p 527.)* Insulin allows the disposition and utilization of glucose—particularly exogenous glucose. Gluconeogenesis is the adaptive response of the organism to low blood levels of glucose and is, therefore, diminished by insulin.

363. The answer is E. *(Williams, ed 5. pp 527-528.)* Among the potential uses for glucose, proteogenesis is as important as lipogenesis and glycogenesis. All these biosynthetic activities are facilitated, not diminished, by insulin.

364. The answer is B. *(White, ed 6. pp 1057-1058, 1277, 1310-1312.)* Somatostatin is produced by the D cells of the pancreatic islets of Langerhans. It is a potent inhibitor of the release of insulin and glucagon, which are produced by the beta and alpha cells of the pancreatic islets respectively. Somatotropin, or growth hormone, is produced by cells of the adenohypophysis and probably regulates the synthesis of some of the growth factors known as somatomedins. Cholecystokinin and secretin, both peptide hormones concerned in the regulation of the digestive system, are produced by cells scattered throughout the intestinal mucosa.

365. **The answer is D.** *(Williams, ed 5. pp 91-92.)* Oxytocin plays only a secondary role in fetal expulsion. Its major role is in the induction of lactation. Suckling, via sensory pathways, in most mammals (but not necessarily in humans) causes oxytocin release, which in turn causes contraction of the mammary alveoli and ductules to force the milk into the larger collecting ducts and cisterns.

366. **The answer is D.** *(Williams, ed 5. pp 854-868.)* Although prostaglandins were originally isolated from prostate glands, seminal vesicles, and semen, their synthesis in other organs has been amply documented; indeed, few organs have failed to demonstrate prostaglandin release.

367. **The answer is D.** *(Williams, ed 5. p 515.)* There are two phases of insulin release during glucose stimulation: an acute release that immediately follows glucose stimulation and depends on readily accessible presynthesized insulin for release; and a secondary phase involving a more protracted release that involves mobilization of stored granules of insulin.

368. **The answer is B.** *(Beeson, ed 14. pp 1368-1374.)* Riboflavin is a light-sensitive yellow compound that is the active coenzyme for several flavoproteins. Manifestations of riboflavin deficiency include corneal vascularization, angular stomatitis, cheilosis, nasolabial seborrhea, and scrotal and vulvar dermatitis. Eye signs of thiamine deficiency are nystagmus and abducens palsy. Avitaminosis A produces nyctalopia and keratomalacia; neither inositol nor niacin deficiency produces characteristic eye signs.

369. **The answer is B.** *(White, ed 6. pp 1265-1268, 1287-1313.)* Oxytocin and vasopressin are both nonapeptides that differ at only two positions: oxytocin has isoleucine and leucine at the 3- and 8-positions, respectively; vasopressin has phenylalanine and arginine at the 3- and 8-positions, respectively. Insulin differs from proinsulin in being much smaller. All the other hormone combinations cited in the question are structurally very dissimilar.

370. **The answer is D.** *(Lynch, ed 2. p 394.)* The glucose tolerance test illustrated in the question shows the typical lag curve of delayed insulin response followed by a hypoglycemic period occurring in two to three hours, or later. This result may be associated with adiposity and early diabetes.

371. **The answer is C.** *(White, ed 6. pp 1247-1263.)* The compound whose structure appears in the question is cholesterol, one of a large group of steroids.

Other members of the general group of steroids derived from cholesterol function as vitamins and hormones.

372. The answer is A. *(White, ed 6. pp 60-62, 1186, 1249.)* Adrenocorticotropin (ACTH) is a peptide hormone of the adenohypophysis that influences the secretion of corticosteroid hormones. The other compounds listed in the question contain the basic steroid-ring system and are ultimate derivatives of cholesterol.

373. The answer is B. *(White, ed 6. pp 1287-1315.)* Vasopressin is elaborated by hypothalamic cells and transported to the posterior pituitary (neurohypophysis) for release. All the other hormones cited in the question are produced by the anterior pituitary (adenohypophysis), and released there.

374. The answer is C. *(White, ed 6. pp 1006-1007.)* Ferrous iron (Fe^{+2}) is the form absorbed in the intestine by ferritin, transported in plasma by transferrin, and stored in the liver in combination with ferritin, or as hemosiderin. There is no known excretory pathway for iron.

375. The answer is C. *(White, ed 6. pp 1322, 1327-1332.)* Certain amino acids and lipids are dietary necessities because humans cannot synthesize them. The energy usually obtained from carbohydrates can be obtained from lipids and the conversion of some amino acids to citric acid cycle intermediates.

376. The answer is E. *(Lehninger, ed 2. pp 495, 546-547.)* Carnitine increases fatty acid transport into mitochondria and thus stimulates fatty acid oxidation. The other compounds listed in the question are members of the transport chain conveying electrons to oxygen in the mitochondrial respiratory chain.

377. The answer is C. *(White, ed 6. pp 1354-1355.)* Pernicious anemia results from an inability to absorb vitamin B_{12} from the gastrointestinal tract. This may be due to a deficiency of intrinsic factor, surgical gastrectomy, or small bowel disease. The earliest clinical signs of pernicious anemia do not appear until three to five years following the onset of vitamin B_{12} deficiency.

378. The answer is B. *(Williams, ed 5. pp 754-755.)* Osteomalacia is the name given to the disease of bone seen in adults with vitamin D deficiency: it is analogous to rickets, which is seen in children with the same deficiency. Both disorders are manifestations of defective bone formation.

379. The answer is D. *(White, ed 6. pp 1360-1361.)* Scurvy results from ascorbic acid (vitamin C) deficiency. It results in failure of mesenchymal cells to form collagen, causing skeletal, dental, and connective tissue deterioration. Anemia in scurvy is associated with impaired utilization of iron, and deranged folate metabolism.

380. The answer is E. *(Stryer, pp 316-320.)* Thiamine pyrophosphate is a coenzyme involved in oxidative decarboxylation of α-keto acids. Other cofactors involved in this type of reaction include FAD, NAD$^+$, lipoic acid, and CoA.

381. The answer is B. *(White, ed 6. pp 1339-1342.)* Pellagra is a disease resulting from deficiency of the vitamin niacin (nicotinic acid). The clinical syndrome characteristic of pellagra, which can include dermatitis, stomatitis, diarrhea, and dementia, actually may result from deficiencies of other nutrients in addition to niacin.

382. The answer is D. *(White, ed 6. pp 1336, 1341-1342, 1354-1355, 1360-1361, 1370.)* Lack of vitamin D, a fat-soluble vitamin involved in calcium metabolism, produces rickets in children. Lack of the water-soluble vitamins thiamine, niacin, B_{12}, and vitamin C, produces beriberi, pellagra, pernicious anemia, and scurvy, respectively.

383. The answer is C. *(White, ed 6. p 1336.)* Both beriberi and Wernicke's disease are thought to result from deficiency of thiamine (vitamin B_1). Birds manifest thiamine deficiency with opisthotonos. In the United States, thiamine deficiency almost invariably is associated with chronic alcoholism and involves intraventricular brain hemorrhage with respiratory and cardiovascular dysfunction.

384. The answer is B. *(Stryer, pp 316-320.)* In the reaction catalyzed by pyruvate dehydrogenase, the thiamine pyrophosphate (TPP) cofactor forms hydroxyethyl TPP with the substrate pyruvate. The hydroxyethyl TPP is then oxidized and transferred to lipoamide, which is formed by the covalent linking of lipoic acid to a lysine side chain amino group of the enzyme dihydrolipoyl transacetylase. Biotin, vitamin A, and vitamin C are not required in oxidative decarboxylation of α-keto acids. NAD$^+$ is required in the enzyme complex to oxidize the reduced flavin dinucleotide prosthetic group of dihydrolipoyl dehydrogenase, but NADP$^+$ is not required.

385. The answer is E. *(Mahler, ed 2. pp 841-842.)* The vitamin folic acid is converted to its cofactor form, tetrahydrofolate, by a two-step reduction catalyzed by folic acid reductase. Methotrexate, an analog of folic acid that binds tightly to the reductase, prevents formation of the cofactor form of the vitamin.

386. The answer is B. *(Harper, ed 16. pp 244-245, 254, 281, 325.)* Pantothenic acid, also called co-acetylase, is a component of CoA. Acetyl CoA is the activated form of acetate employed in acetylation reactions, including the citric acid cycle and lipid and cholesterol metabolism.

387. The answer is B. *(White, ed 6. pp 1295-1308.)* Follicle-stimulating hormone, luteinizing hormone, human chorionic gonadotropin, and thyroid-stimulating hormone all are glycoproteins composed of two subunits. One of the subunits is common to, or very similar in, all four hormones. ACTH is a single-chain polypeptide of 39 amino acids, and has no carbohydrate moiety.

388. The answer is A. *(Stryer, pp 316-320.)* Pyruvate dehydrogenase is an organized enzyme assembly containing pyruvate dehydrogenase, dihydrolipoyl transacetylase, dihydrolipoyl dehydrogenase, and two enzymes involved in regulation of the overall enzymatic activity of the complex. Pyruvate dehydrogenase requires thiamine pyrophosphate as its cofactor, dihydrolipoyl transacetylase requires lipoic acid and coenzyme A (CoA), and dihydrolipoyl dehydrogenase has an FAD prosthetic group which is reoxidized by NAD^+. The only cofactor listed in the question that is not required in the pyruvate dehydrogenase reaction is pyridoxal phosphate, a cofactor widely used by enzymes involved in amino acid metabolism, like the transaminases.

389. The answer is E. *(White, ed 6. pp 1348-1349.)* Raw egg white contains a protein, avidin, which complexes specifically with biotin and has been shown to induce biotin deficiency in laboratory animals. It was by means of avidin that the factor biotin was first shown to be a growth factor.

390. The answer is B. *(White, ed 6. pp 586-587, 1348.)* The vitamin biotin is the cofactor required by carboxylating enzymes such as acetyl CoA, pyruvate, and propionyl CoA carboxylases. The fixation of CO_2 by these biotin-dependent enzymes occurs in two stages. In the first, bicarbonate ion reacts with ATP and the biotin carrier protein moiety of the enzyme, and in the second stage, the "active CO_2" reacts with the substrate—e.g., acetyl CoA.

391. The answer is A. *(White, ed 6. pp 1346-1348.)* Biotin is a cofactor of many carboxylase enzymes. Succinic thiokinase, however, is not a carboxylase, does not require biotin and, therefore, is not affected by avidin.

392. The answer is D. *(White, ed 6. pp 657, 1343-1345.)* Pyridoxal phosphate is an obligatory cofactor for the reactions of transaminases, and of many other enzymes with amino acid substrates. In these reactions, the vitamin forms a Schiff base intermediate which then can undergo such modifications as rearrangement and transamination.

393. The answer is C. *(White, ed 6. pp 1345-1347, 1349, 1352-1354.)* Folic acid, which contains a pterin ring that cannot be synthesized by humans, is therefore, an essential dietary factor. Bacterial synthesis of the pteridine nucleus uses GTP as starting material. Para-aminobenzoate and glutamate also are required. Vitamin B_{12} contains the corrin ring, which humans cannot synthesize, and cyanocobalamin is the form of vitamin B_{12} extracted from bacterial sources. Biotin is a vitamin synthesized by bacteria and molds, and CoA (Coenzyme A) contains the vitamin pantothenic acid which is synthesized from α-ketovaleric acid by green plants and most microorganisms.

394. The answer is A. *(White, ed 6. pp 1341-1342.)* Pellagra is thought to result chiefly from a deficiency of nicotinic acid (niacin). Photodermatitis, stomatitis, glossitis, and diarrhea, along with central neurologic changes, are found in pellagra; all these symptoms may be alleviated by niacin.

395. The answer is E. *(White, ed 6. pp 1337-1339.)* Deficiency of riboflavin may cause fissuring at the corners of the mouth and lips (cheilosis). (Cheilosis is more commonly seen with iron deficiency and with dentures.) Riboflavin deficiency also is associated with corneal vascularization, seborrheic dermatitis, and glossitis.

396. The answer is B. *(White, ed 6. pp 1345-1346.)* Pantothenate is the precursor of coenzyme A, and together they partake in numerous reactions throughout the metabolic scheme. There is no documented deficiency state for pantothenate.

397. The answer is C. *(White, ed 6. pp 1371-1373.)* Vitamin E is a tocopherol. Deficiency of this fat-soluble vitamin in some species of animals is associated with gonadal dysfunction, anemia, and skin changes. Although vitamin E deficiency occurs in patients with fat malabsorption, the biochemical consequences of hypovitaminosis E in humans remain to be clarified.

398. The answer is B (1, 3). *(White, ed 6. pp 1276-1277.)* Glucagon accelerates glycogenolysis in the liver by activating phosphorylase kinase and, consequently, liver phosphorylase. Skeletal muscle is not a target for glucagon action. Gluconeogenesis is accelerated, not depressed, by glucagon.

399. The answer is B (1, 3). *(Mahler, ed 2. pp 47-50, 66, 635.)* Prostaglandins are derived from homo-γ-linolenic acid (a fatty acid). Vasopressin is a nonapeptide secreted by the pituitary. Thyroxine and epinephrine are derived from tyrosine.

400. The answer is A (1, 2, 3). *(Williams. ed 5. pp 40-46.)* Human chorionic gonadotropin (HCG), LH, TSH, and FSH all are glycoprotein hormones containing two polypeptide chains. The alpha chains are common to all four hormones, whereas the beta chains differ.

401. The answer is B (1, 3). *(Mahler, ed 2. pp 168-170.)* Insulin, which consists of two different polypeptide chains held together by disulfide bonds, is synthesized by cleaving a "connecting peptide" out of a larger molecule called proinsulin. When proinsulin is denatured, it can be renatured to give native molecules because the primary structure of the polypeptide chain determines its higher-order structure. In contrast, because of the absence of the connecting peptide, the primary structure of insulin does not determine its higher-order structure. After denaturation, insulin cannot be renatured successfully in good yield.

402. The answer is E (all). *(Mahler, ed 2. pp 513, 545, 803, 894.)* The vitamin whose structure appears in the question is nicotinic acid. NAD^+ is a cofactor required by lactate dehydrogenase, UDP glucose epimerase, and polynucleotide ligase of *E. coli*. NADPH is a cofactor for phenylalanine hydroxylase.

403. The answer is A (1, 2, 3). *(Mahler, ed 2. pp 401-406.)* The vitamin whose structure appears in the question is thiamine. Thiamine pyrophosphate is required for the reactions catalyzed by pyruvate dehydrogenase, transketolase, and α-ketoglutarate dehydrogenase. In all these reactions, thiamine is involved with oxidative decarboxylation. Glutamate dehydrogenase involves neither oxidative decarboxylation nor thiamine.

404. The answer is A (1, 2, 3). *(White, ed 6. pp 483-485.)* The structure shown in the question is that of the vitamin pyridoxine (vitamin B_6). The active cofactor derived from pyridoxine is pyridoxal phosphate. All transaminase enzymes, and many other enzymes involved in amino acid metabolism, require pyridoxal phosphate as a cofactor. Thus, both glutamic-oxaloacetic transaminase and dihydroxyphenylalanine decarboxylase require pyridoxal phosphate. The cofactor is also essential for the catalytic activity of glycogen phosphorylase in which it appears to play a purely structural role.

405. The answer is C (2, 4). *(Mahler, ed 2. pp 414-418, 519-520.)* The structure shown in the question is that of the vitamin riboflavin. It is a precursor of two cofactors involved in electron transport systems, riboflavin 5′-phosphate, also known as flavin mononucleotide (FMN), and flavin adenine dinucleotide (FAD). Strictly speaking, these compounds are not nucleotides, as they contain the sugar alcohol ribitol, not ribose. The cofactors are strongly bound to their apoenzymes and function as dehydrogenation catalysts. Pyruvate dehydrogenase is a multienzyme complex and contains the enzyme dihydrolipoyl dehydrogenase which has as its prosthetic group two molecules of FAD per molecule of enzyme. In the overall reaction, the reduced FAD is reoxidized by NAD^+. Succinate dehydrogenase also contains tightly bound FAD, one molecule per molecule of enzyme. Glutamate and lactate dehydrogenases both use nicotinamide dinucleotide cofactors and do not contain FAD as a prosthetic group.

406. The answer is D (4). *(Stryer, pp 321, 357-361. White, ed 6. pp 445-446.)* The only enzyme listed in the question that requires thiamine pyrophosphate as a cofactor is transketolase. Erythrocyte levels of transketolase have been found to be a sensitive indicator of thiamine deficiency. Glucose 6-phosphate dehydrogenase requires nicotinamide adenine dinucleotide phosphate (NADP) as a cofactor, and 2,3-diphosphoglycerate mutase has a requirement for catalytic amounts of 2,3-diphosphoglycerate. Transaldolase does not have a cofactor requirement.

407. The answer is A (1, 2, 3). *(Metzler, p 736. White, ed 6. pp 1358, 1363-1364, 1367-1371.)* Retinol (vitamin A), cholecalciferol (vitamin D_3), and tocopherol (vitamin E) are lipid soluble compounds that can be regarded as composed of isoprene units $(CH_2=C(CH_3)CH=CH_2)$. Retinol and tocopherol cannot be synthesized by higher animals and have to be supplied in the diet (they are true vitamins). Cholecalciferol is formed in the skin of humans by the action of sunlight on a cholesterol precursor, 7-dehydrocholesterol. Cholecalciferol is the precursor of the physiologically active forms of vitamin D, 1,25-dihydroxycholecalciferol and 1,24,25-trihydroxycholecalciferol. Ascorbic acid is a water-soluble compound and an essential food factor in humans (vitamin C).

408. The answer is E (all). *(Mahler, ed 2. pp 393-401.)* Pyridoxal phosphate is a cofactor in many types of reactions: decarboxylation (glutamate decarboxylase); deamination (serine dehydratase); transamination (glutamic-oxaloacetic transaminase); and transsulfuration (cystathionine synthetase and cystathionase).

409. The answer is B (1, 3). *(White, ed 6. pp 1211, 1248, 1294-1295.)* Parathormone is secreted by the parathyroid glands and is involved in calcium and phosphate homeostasis. Aldosterone is a mineralocorticoid produced by the adrenal gland. Only ACTH and FSH are produced by the adenohypophysis.

410. The answer is D (4). *(White, ed 6. pp 1352-1357.)* Vitamin B_{12}, a complex molecule containing cobalt and a porphyrin-like ring system, is a water-soluble vitamin. The B vitamins are generally water-soluble, whereas vitamins A, D, E, and K are all fat-soluble.

411. The answer is A (1, 2, 3). *(White, ed 6. pp 325, 328, 373.)* FAD, CoA, and NAD contain, respectively, riboflavin, pantothenic acid, and nicotinic acid, all of which are vitamins. ATP is composed of adenosine, ribose, and phosphate only.

412. The answer is B (1, 3). *(Mahler, ed 2. pp 418-422, 424-427, 656-657.)* Hemoglobin and cytochrome *c* contain heme cofactors that are porphyrin derivatives. Methylmalonyl CoA mutase requires vitamin B_{12} which contains a corrin ring resembling, but different from, a porphyrin ring. Ferredoxin contains nonheme iron.

413. The answer is B (1, 3). *(Lehninger, ed 2. pp 345-347, 729-737.)* The vitamin whose structure appears in the question is folic acid. Tetrahydrofolic acid, the active cofactor derived from folic acid, is required in two steps of purine synthesis, and thus required in the *de novo* synthesis of ATP and GTP. CTP and TTP are pyrimidine base derivatives and, while *de novo* synthesis of the pyrimidine ring does not require tetrahydrofolate, the formation of thymine from uracil does.

414. The answer is A (1, 2, 3). *(Williams, ed 5. p 504.)* The six-amino acid ring characteristic of insulin, oxytocin, and vasopressin may be a structural property of the molecule that is involved in binding to specific receptor sites on target cells.

415. The answer is D (4). *(Sawin, pp 4-6.)* Since actinomycin D blockage of RNA synthesis, and mitomycin C blockage of DNA snythesis, failed to inhibit the action of the hormone mentioned in the question, DNA synthesis and RNA synthesis were not required. Puromycin and cycloheximide blocked the hormone through their action as inhibitors of protein synthesis.

416-418. The answers are: 416-A, 417-C, 418-E. *(White, ed 6. pp 1186, 1211, 1241.)* ACTH, a peptide hormone produced by the adenohypophysis, stimulates the synthesis of steroids by the adrenal cortex. The neurohypophysis (posterior pituitary) produces vasopressin and oxytocin. Both epinephrine and norepinephrine are produced in the adrenal medulla; steroid hormones are produced by the adrenal cortex. The parafollicular cells of the thyroid gland produce calcitonin, which regulates calcium metabolism; these are the cells in which medullary carcinoma of the thyroid originates. Calcitonin assay provides a specific means of following such a tumor's course.

419-421. The answers are: 419-E, 420-A, 421-B. *(White, ed 6. pp 1049-1057.)* Secretin, a circulatory hormone liberated in response to peptides or acid in the duodenum, stimulates the flow of pancreatic juice. Gastrin governs acid production by the stomach, and cholecystokinin causes the gallbladder to contract. Cholecystokinin stimulates this contraction after it is released by the duodenum into the circulation, with subsequent emptying of bile into the intestine. The C-terminal octapeptide of cholecystokinin is more than five times as potent as the parent hormone, and its C-terminal pentapeptide is identical with gastrin. Gastrin, produced in specialized cells of the antral mucosa of the stomach, stimulates parietal cells to produce HCl (approximately 0.16 M) and KCl (0.007 M); it also stimulates glucagon and insulin secretion. Gastrin production is inhibited by secretin.

422-424. The answers are: 422-B, 423-C, 424-E. *(White, ed 6. pp 1044-1050.)* Pancreatic juice, which neutralizes the acidity of gastric secretions, contains approximately 80 mEq/l of HCO_3^-; it also contains appreciable quantities of potassium. The parietal cells of the stomach, which secrete 0.16 M HCl and traces of other electrolytes, are stimulated by both neural (vagal) and hormonal (gastrin) mechanisms. Stimulation by exogenous histamine provides a routine clinical test of gastric acid secretion. Tears contain the enzyme lysozyme that can protect the eye from infection by hydrolyzing the cell walls of many bacteria.

425-427. The answers are: 425-A, 426-C, 427-D. *(Metzler, pp 735-737.)* Cholecalciferol is formed in the skin by the action of sunlight on the $\Delta^{5,7}$-unsaturated sterol, 7-dehydrocholesterol. This sterol is a precursor of cholesterol and is found in high concentration in skin cells. 25-Hydroxycholecalciferol is formed from cholecalciferol in the liver. To a certain extent, its rate of production is regulated by strong product inhibition of the enzyme system. 1,25-Dihydroxycholecalciferol, which is one of the physiologically active forms of vitamin D, is formed from 25-hydroxycholecalciferol in the kidney. The 1,25-dihydroxy derivative has direct effects on calcium transport in intestinal mucosa and in bone cells. A trihydroxy derivative (1,24,25), also formed in the kidney, has a specific effect on calcium uptake by the small intestine.

Membranes and Cell Structure

DIRECTIONS: Each question below contains five suggested answers. Choose the **one best** response to each questions.

428. A typical plasma membrane would be most likely to have which of the following weight compositions?

	Lipid	Protein	Carbohydrate	RNA
(A)	35%	45%	5%	10%
(B)	35%	55%	5%	0%
(C)	20%	75%	0%	0%
(D)	60%	30%	0%	5%
(E)	35%	40%	20%	0%

429. Which of the following compounds is an inhibitor of sodium-dependent glucose transport across the plasma membrane?

(A) Ouabain
(B) Sodium azide
(C) Dicumarol
(D) Phlorhizin
(E) Phloretin

430. Which of the following enzymes, or enzyme systems, is localized in the inner membrane of the mitochondrion?

(A) Acyl-CoA synthetases
(B) Isocitrate dehydrogenase
(C) Fatty acyl-CoA oxidation enzymes
(D) Succinate dehydrogenase
(E) Nucleoside diphosphate kinase

Questions 431-432

431. What is the structure indicated in the electron micrograph below?

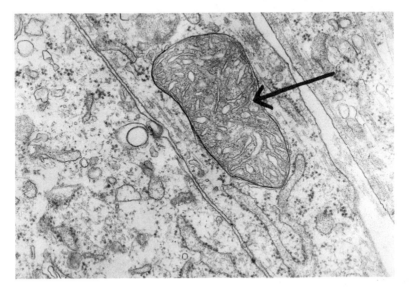

(A) Golgi apparatus
(B) Lysosome
(●) Mitochondrion
(D) Nucleus
(E) Rough endoplasmic reticulum

432. The structure indicated in the electron micrograph is involved primarily in

(●) energy production
(B) enzymatic degradation of ingested particles
(C) packaging of secretory proteins
(D) protein synthesis
(E) transmission of genetic information

433. The electron micrograph below shows a protein structure found in mitotic spindles, eukaryotic flagella, and nerve axons. What is the structure?

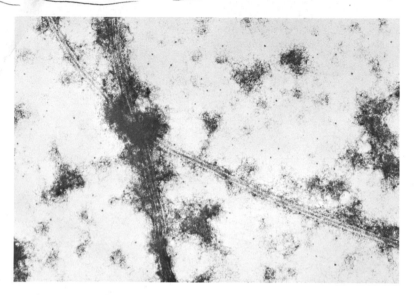

(A) Ribosomes
(B) Actinomycin
(C) Microtubules
(D) Endoplasmic reticulum
(E) Microfilaments

434. The membrane bridge illustrated in the electron micrograph below is a

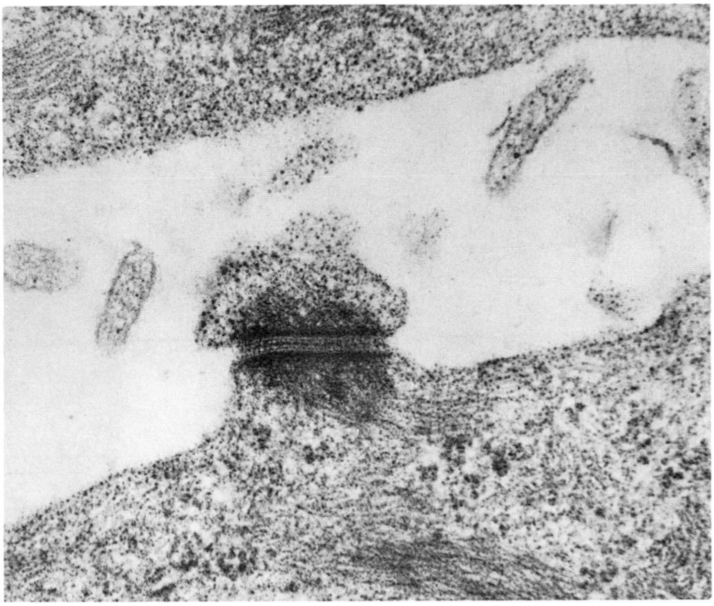

(A) septate junction
(B) gap junction
(C) conjugation junction
(D) zonula occludens
(E) desmosome

DIRECTIONS: Each question below contains four suggested answers of which **one** or **more** is correct. Choose the answer:

A	if	**1, 2, and 3**	are correct
B	if	**1 and 3**	are correct
C	if	**2 and 4**	are correct
D	if	**4**	is correct
E	if	**1, 2, 3, and 4**	are correct

435. Chain elongation of fatty acids occurring in mammalian liver takes place in which of the following subcellular fractions?

(1) Nuclear
(2) Mitochondrial
(3) Microsomal
(4) Supernatants of 100,000 x g
     (60 min)

436. Crucial features of the unit-membrane model of Davson and Danielli include

(1) a bimolecular leaflet of lipids
     (bilayer)
(2) a hydrophobic interior and
     hydrophilic exterior
(3) a monolayer of protein at the
     aqueous interfaces of the mem-
     brane with its environment
(4) cholesterol molecules inter-
     spersed in the membrane to
     increase stability

437. The site of intracellular protein synthesis is the

(1) rough endoplasmic reticulum
(2) Golgi apparatus
(3) polysomes
(4) nucleus

438. Which of the following enzymes are found in lysosomes?

(1) Ribonuclease
(2) β-Galactosidase
(3) Acid phosphatase
(4) Cathepsins

439. The structure indicated in the electron micrograph below is

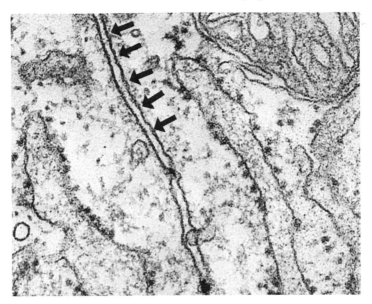

(1) composed of protein and lipid bilayers
Ɓ(2) void of enzymatic activity
(3) capable of transporting Na^+ against a chemical gradient
(4) the primary producer of intra-cellular energy

SUMMARY OF DIRECTIONS

A	B	C	D	E
1, 2, 3 only	1, 3 only	2, 4 only	4 only	All are correct

440. Red blood cell ghosts shown in the electron micrograph below may correctly be said to

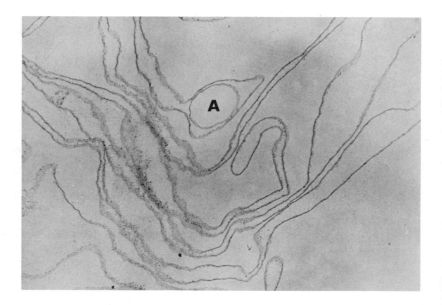

(1) be devoid of cholesterol
(2) have lost their membrane bilayer
(3) contain internal mitochondrial
    vesicles (A)
(4) have lost mainly hemoglobin

DIRECTIONS: The groups of questions below consist of lettered choices followed by several numbered items. For each numbered item select the **one** lettered choice with which it is **most** closely associated. Each lettered choice may be used once, more than once, or not at all.

Questions 441-444

For each cellular structural element listed below, select the appropriate lettered location in the microphotograph. Answer E if the structure does not correspond to any lettered item.

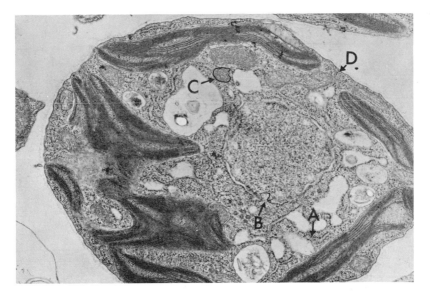

Å 441. Endoplasmic reticulum

Ƀ 442. Nuclear pore

Ŀ 443. Mitochondrion

Ɗ 444. Plasma membrane

Questions 445-447

For each biochemical process that follows, select the lettered area in the electron micrograph below with which it is associated. Answer E if the process is not associated with any of the designated areas.

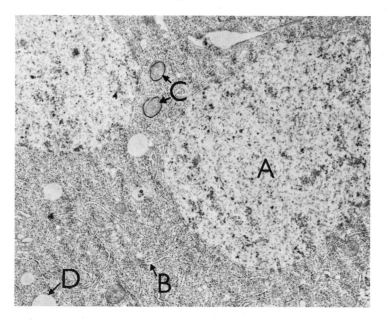

A 445. RNA synthesis

C 446. Oxidative phosphorylation

B 447. Protein synthesis

Membranes and Cell Structure

Answers

428. The answer is B. *(Mahler, ed 2. pp 457-458.)* A typical plasma membrane contains 35-40 percent lipid, 55 percent protein, up to 5 percent carbohydrate, and less than 0.1 percent RNA by weight.

429. The answer is D. *(White, ed 6. pp 33, 426-427.)* Phlorhizin is an inhibitor of sodium-dependent glucose transport, while phloretin inhibits sodium-independent facilitated diffusion. Glycosides both, they are related to the digitalis glycoside ouabain, which inhibits Na⁺-K⁺-ATPase which ejects Na⁺ from cells.

430. The answer is D. *(Lehninger, ed 2. p 512.)* The mitochondrion can be considered to have several "compartments"—the outer membrane, the space between inner and outer membranes, the inner membrane, and the matrix. Acyl-CoA synthetases and nucleotide diphosphate kinase are localized in the outer membrane, succinate dehydrogenase in the inner membrane; fatty acyl-CoA oxidation enzymes and isocitrate dehydrogenase dwell in the matrix.

431. The answer is C. *(Mahler, ed 2. pp 433-434.)* The structure shown in the question is a mitochondrion, characterized by a smooth outer membrane and a complex inner membrane with many folds called cristae. This organelle contains the major energy-generating system in eukaryotic cells—the coupled electron transport-oxidative phosphorylation system.

432. The answer is A. *(Mahler, ed 2. pp 433-434.)* The structure shown in the question is a mitochondrion, which is the energy-producing organelle of the cell. Lysosomes are the degradative machinery of the cell, and Golgi bodies are engaged in the packaging of secretory proteins. Ribosomes accomplish protein systhesis and DNA accomplishes eukaryotic cell genetic transfer.

433. The answer is C. *(Wolfe, pp 346-347.)* Microtubules, as shown in the question, are hollow, tubular structures 100Å in diameter composed of two protein subunits. They have a structural function in the cell, and are associated with the movement of chromosomes and the flow of neurotransmitters in axons.

434. The answer is E. *(Wolfe, p 20.)* The electron micrograph appearing in the question exhibits a membrane bridge known as a desmosome. Desmosomes are dense plaques consisting of specialized fibrils found between epithelial cells; they serve to connect the membranes of adjacent cells.

435. The answer is C (2, 4). *(White, ed 6. p 591.)* The elongation of medium-chain fatty acids (around twelve carbon atoms) occurs most readily in the mitochondria and does not require malonyl CoA. However, the elongation of poly-unsaturated fatty acids, which occurs in the soluble fraction, does utilize malonyl CoA.

436. The answer is A (1, 2, 3). *(Wolfe, pp 74-75.)* Crucial features of the Davson-Danielli model of membrane structure include an arrangement of membrane lipids in a bilayer with hydrophobic side chains pointing inward. The hydrophilic heads, which are directed outward, interact with a monolayer of protein which in turn interacts with the aqueous environment.

437. The answer is B (1, 3). *(Wolfe, p 272.)* Protein synthesis occurs in the cytoplasm, on groups of ribosomes called polysomes, and on ribosomes associated with membranes, termed the rough endoplasmic reticulum. The Golgi apparatus is involved in the secretion of macromolecules, such as proteins, by cells.

438. The answer is E (all). *(White, ed 6. pp 311, 425, 703.)* Lysosomes are cellular organelles containing hydrolytic enzymes that degrade macromolecules taken up by these organelles. The enzymes include ribonuclease, which hydrolyses RNA, β-galactosidase which hydrolyzes oligosaccharides at an acid pH, acid phosphatase which attacks phosphate esters (also at an acid pH), and cathepsins which are intracellular proteolytic enzymes that hydrolyze peptide bonds.

439. The answer is B (1, 3). *(Wolfe, p 15.)* The structure shown in the electron micrograph accompanying the question is a plasma membrane composed of protein and lipid bilayers. The membrane contains a Na^+/K^+ pump that is dependent on an ATPase enzyme. Cellular energy is produced primarily in mitochondria.

440. The answer is D (4). *(Bloom, ed 10. p 138.)* Lysed red blood cell ghosts contain little hemoglobin. They are well-studied preparations of intact membrane bilayers with the usual content of cholesterol, and no mitochondria.

441-444. The answers are: 441-A, 442-B, 443-C, 444-D. *(Bloom, ed 10. pp 45-48, 112-113. Wolfe, p 16.)* The subcellular organelles exhibited in the question are: A) a bilayer membrane with or without attached ribosomes in the cytosol; B) a continuity between the inner and outer nuclear membranes; C) a membrane-bound body containing cristae, a finely granular matrix, and elementary bodies; D) the limiting bilayer membrane of the cell.

445-447. The answers are: 445-A, 446-C, 447-B. *(Mahler, ed 2. pp 432-438.)* The nucleus is the site of both DNA replication and RNA transcription. DNA remains in the nucleus, while mRNA is transported to the cytoplasm to direct protein synthesis. The mitochondria are the sites of the electron transport chain that is coupled to the phosphorylation of ADP to ATP, which ultimately oxidizes molecular oxygen to water. The endoplasmic reticulum consists of lipoprotein membranes, and is divided into smooth and rough types. The smooth endoplasmic reticulum lacks ribosomes; the rough endoplasmic reticulum is lined with ribosomes and accomplishes protein synthesis.

✓ Metabolism

DIRECTIONS: Each question below contains five suggested answers. Choose the **one best** response to each questions.

448. Which of the values listed below most closely approximates the Respiratory Quotient (RQ) of an animal oxidizing fat?

(A) 2.0
(B) 1.0
(C) 0.85
(●) 0.7
(E) 0.5

449. The oxygen dissociation curve for hemoglobin is shifted to the right by

(A) decreased O_2 tension
(B) increased N_2 tension
(C) increased pH
(●) increased CO_2 tension
(E) decreased CO_2 tension

450. A .22 M solution of lactic acid (pKa 3.9) was found to contain .20 M in the dissociated form, and .02 M undissociated. What is the pH of the solution?

(A) pH 2.9
(B) pH 3.3
(C) pH 3.9
(●) pH 4.9
(E) pH 5.4

451. Which of the lettered points on the diagram below best represents completely compensated metabolic acidosis?

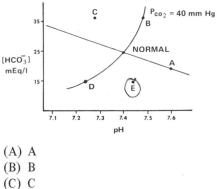

(A) A
(B) B
(C) C
(D) D
(●) E

452. The component in muscle tissue that contains the ATPase activity required for contraction is

(A) actin
(●) myosin
(C) sarcoplasmic reticulum
(D) motor end-plate
(E) calcium

453. A healthy (70-kg) man eats a well-balanced diet containing adequate calories and 62.5 g of high quality protein per day. Measured in grams of nitrogen, his daily nitrogen balance would be

(A) +10 g
(B) +6.25 g
(●) 0 g
(D) −6.25 g
(E) −10 g

454. Addition of which of the following substances to a cell-free cytoplasmic suspension would decrease levels of cyclic AMP?

(●) Cyclic AMP phosphodiesterase
(B) Dibutyryl cyclic AMP
(C) Caffeine
(D) Aminophylline
(E) Adenylate cyclase

455. The majority of energy for muscular contraction is stored in muscle tissue in the form of

(A) ADP
(B) phosphoenolpyruvate
(C) cyclic AMP
(D) ATP
(●) creatine phosphate

456. The pH of body fluids is stabilized by buffer systems. Which of the following compounds is the most effective buffer at physiologic pH?

(A) $Na_2 HPO_4$, pKa = 12.32
(B) $NH_4 OH$,   pKa = 9.24
(●) $NaH_2 PO_4$, pKa = 7.21
(D) $CH_3 CO_2 H$, pKa = 4.74
(E) Citric acid, pKa = 3.09

457. Water, which comprises 70 percent of body weight, may be said to be the "cell solvent." The property of water that most contributes to its ability to dissolve compounds is the

(A) strong covalent bonds formed between water and salts
(●) hydrogen bonds formed between water and biochemical molecules
(C) hydrophobic bonds formed between water and long chain fatty acids
(D) absence of interacting forces
(E) fact that the freezing point of water is much lower than body temperature

458. The normal osmotic pressure of proteins in the blood is approximately

(A)   0 mm Hg
(B) 10 mm Hg
(●) 30 mm Hg
(D) 70 mm Hg
(E) 90 mm Hg

459. The principal osmotic regulatory component of plasma is

(A) immunoglobulin
(B) erythrocytes
(C) urea
(D) albumin
(E) glucose

460. Consider a suspension of mitochondria incubated with a respiratory substrate, e.g., succinate, plus phosphate. Concerning this suspension, all the following statements regarding oxygen consumption would be true EXCEPT that it would

(A) be increased by the addition of ADP
(B) be increased by oligomycin if 2,4-dinitrophenol were present
(C) be increased by 2,4-dinitrophenol if oligomycin were present
(D) not be increased by ADP if oligomycin were present
(E) not be increased by ADP if 2,4-dinitrophenol were present

461. The plasma acid phosphatase level may be significantly elevated in

(A) metastatic prostatic carcinoma
(B) osteoblastic sarcoma
(C) obstructive jaundice
(D) Paget's disease
(E) rickets

462. Which of the schematic configurations below represents a stable lipid-water interaction?

LIPID MOLECULE

POLAR HEAD

NONPOLAR TAIL

A

B

AIR

C

D

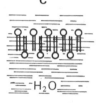

E

(A) Figure A
(B) Figure B
(C) Figure C
(D) Figure D
(E) Figure E

463. Which of the following biochemical phenomena is charateristic only of mammalian liver?

(A) Gluconeogenesis
(B) Glycogen synthesis and storage
(C) Serum albumin synthesis
(D) Hydroxylation of phenylalanine
(E) Epinephrine sensitivity

464. Which of the following blood proteins is the major source of extracellular cholesterol for the tissues in humans?

(A) Very low-density lipoprotein
(B) Low-density lipoprotein
(C) High-density lipoprotein
(D) Albumin
(E) γ-Globulin

465. Sandhoff's gangliosidosis is characterized by a deficiency of

(A) hexosaminidase A, only
(B) hexosaminidase B, only
(C) hexosaminidase C, only
(D) hexosaminidase A and B
(E) hexosaminidase A, B, and C

466. Which of the following is a sex-linked recessive disease?

(A) Lesch-Nyhan syndrome
(B) Cushing's syndrome
(C) Maple syrup urine disease
(D) Familial hypercholesterolemia
(E) Infantile autism

467. When fresh urine contains large amounts of porphobilinogen, its color is

(A) normal
(B) pink to light red
(C) green
(D) dark red to brown
(E) black

468. A theoretical nonequilibrium situation involving a membrane permeable only to sodium and chloride is shown below. According to the Donnan equilibrium, what will be the final concentration of chloride on the left?

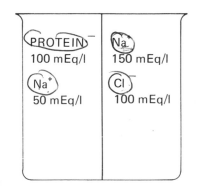

(A) 0
(B) 25 mEq/1
(C) 33 mEq/l
(D) 50 mEq/l
(E) 75 mEq/l

469. McArdle's syndrome (a glycogen storage disease) involves a deficiency in which of the following enzymes?

(A) Pancreatic peptidase
(B) Hepatic glycogen synthetase
(C) Hepatic phosphorylase
(D) Muscle phosphorylase
(E) Debranching enzyme

470. The enzyme secreted by osteo-blasts that is thought to provide phosphate for bone deposition is

(A) glucose 6-phosphate dehydrogenase
(B) acid phosphatase
(O) alkaline phosphatase
(D) adenosine triphosphatase
(E) phosphorylase kinase

471. Phenylketonuria is caused by a lack of

(A) phenylalanine hydroxylase
(B) phenylalanine α-ketoglutaric transaminase
(C) homogentisate oxidase
(D) dopa decarboxylase
(E) cystathionine synthetase

472. Which of the following enzymes is deficient in von Gierke's disease?

(A) Uridine diphosphate glucose pyrophosphorylase
(B) Glucose 6-phosphatase
(C) α-1,6-Glucosidase
(D) Glucokinase
(E) Phosphoglucomutase

473. Which of the following diseases is NOT caused by a functionally defective enzyme?

(A) Phenylketonuria
(B) Tay-Sachs disease
(C) Von Gierke's disease
(D) Gaucher's disease
(E) Caisson disease

474. In type IV hereditary disorders of glycogen metabolism (Andersen's disease), an abnormal glycogen with few branches is formed and stored in the liver. This condition results from an absence of which of the following enzymes?

(A) Glucose 6-phosphatase
(B) Amylo-1,6-glucosidase
(C) Amylo-(1,4→1,6)-transglycosylase
(D) Glycogen phosphorylase
(E) Phosphorylase kinase

475. Of the following body fluids, the one with the lowest pH is

(A) plasma
(B) pancreatic juice
(C) liver bile
(D) gastric juice
(E) sweat

DIRECTIONS: Each question below contains four suggested answers of which **one** or **more** is correct. Choose the answer:

A	if	1, 2, and 3	are correct
B	if	1 and 3	are correct
C	if	2 and 4	are correct
D	if	4	is correct
E	if	1, 2, 3, and 4	are correct

476. Which of the following statements about the sequence and control of metabolic functions are true?

(1) The anabolic pathway of any given molecule usually is the reverse of its catabolic pathway
(2) Anabolism allows the synthesis of macromolecules from small precursors and has definite energy requirements
(3) A given set of enzymes generally is exclusive to one metabolic pathway and thus allows for independent control mechanisms
(4) The initial reactions of most metabolic sequences are thermodynamically irreversible and frequently determine metabolic control

477. Which of the following substances decrease the affinity of hemoglobin for oxygen?

(1) 2,3-Diphosphoglycerate
(2) Carbon dioxide
(3) Hydrogen ions
(4) Hydroxyl ions

478. Starvation induces an increase in activity of the hepatic enzymes involved in

(1) the hexose monophosphate shunt
(2) lipogenesis
(3) glycolysis
(4) gluconeogenesis

479. Mammalian tissue is able to synthesize which of the following compounds?

(1) Biotin
(2) Choline
(3) Dehydroshikimic acid
(4) Inositol

480. Damage to which of the following tissue cells elevates serum alkaline phosphatase levels?

(1) Biliary tract
(2) Osteoblasts
(3) Small intestine mucosa
(4) Placenta

481. Which of the following structures are able to oxidize glucose?

(1) Liver
(2) Brain
(3) Heart
(4) Erythrocytes

SUMMARY OF DIRECTIONS

A	B	C	D	E
1, 2, 3 only	1, 3 only	2, 4 only	4 only	All are correct

482. Which of the following structures are capable of gluconeogenesis?

(1) Erythrocytes
(2) Kidney
C (3) Skeletal muscle
(4) Liver

483. A patient with tyrosinosis suffers from which of the following biochemical abnormalities?

(1) Excretion of para-hydroxyphenyl-pyruvic acid
B (2) Inability to utilize homogentisic acid
(3) Absence of an oxidase
(4) Elevated levels of phenyllactic acid

484. Phenylketonuria is thought to be a genetic disease related to

(1) a constitutive mutation in the synthesis of shikimic acid
(2) tyrosinase deficiency
(3) deficiency of a phenylalanine transport enzyme
D (4) deficiency of a hydroxylase in an amino acid conversion

485. In maple syrup urine disease, the α-ketoacids that accumulate are derived from which of the following amino acids?

(1) Leucine
B (2) Lysine
(3) Isoleucine
(4) Phenylalanine

486. Urine that darkens on standing may contain

(1) metanephrine
(2) porphobilinogen
(3) 5-hydroxyindoleacetic acid
C (4) homogentisic acid

487. The inability to utilize glycogen as a source of glucose is associated with which of the following enzyme deficiencies?

(1) Amylo-1,6-glucosidase
A (2) Phosphorylase
(3) Glucose 6-phosphatase
(4) Amylo-(1,4→1,6)-transglycosylase

488. An abnormal glucose tolerance test may occur in patients suffering from

(1) dumping syndrome
(2) rheumatoid arthritis
(3) oat-cell carcinoma
E (4) diabetes

489. Once a diabetic patient is placed on chronic insulin therapy,

(1) serum insulin levels can no longer be measured accurately
(2) impotence occurs within a few months
(3) insulin requirements may decrease to nothing for a period of time before increasing
(4) a diet without any restrictions may be completely tolerable

490. Heavy alcohol ingestion may be associated with

(1) fatty liver change
(2) polyneuropathy
(3) ketoacidosis
(4) Wernicke-Korsakoff syndrome

491. Which of the following structures are capable of oxidizing ketone bodies to carbon dioxide?

(1) Erythrocytes
(2) Brain
(3) Liver
(4) Heart

492. Which of the following structures are capable of synthesizing ketone bodies from fatty acids?

(1) Erythrocytes
(2) Brain
(3) Skeletal muscle
(4) Liver

164

DIRECTIONS: The groups of questions below consist of lettered choices followed by several numbered items. For each numbered item select the **one** lettered choice with which it is **most** closely associated. Each lettered choice may be used once, more than once, or not at all.

Questions 493-496

For each enzyme that follows, select the hereditary disorder most likely to result from its deficiency.

(A) Hartnup's disease
(B) Albinism
(C) Crigler-Najjar syndrome
(D) Hemolytic anemia
(E) Wilson's disease

A 493. Tryptophan pyrrolase A

C 494. Uridine diphosphate glucuronate transferase C

E 495. Ceruloplasmin e

D 496. Pyruvate kinase d

B — Tyrosinase - B.

Questions 497-500

For each biochemical description that follows, select the diagnosis or diagnoses to which it is most appropriate.

(A) Juvenile diabetes
(B) Adult onset diabetes
(C) Both
(D) Neither

A 497. Absolute need for insulin A

A 498. Frequent ketoacidosis A

A 499. Unstable and "brittle" diabetes A

D 500. Insulin resistance d

Metabolism

Answers

448. The answer is D. *(White, ed 6. p 294.)* The consumption of oxygen and production of CO_2 are consequences of the ultimate oxidation of metabolic fuels. The CO_2-to-O_2 ratio, known as the respiratory quotient (RQ), can be calculated for fat, carbohydrate, and protein. When fat is the predominant fuel oxidized, the RQ approaches 0.7. If carbohydrate is the major fuel, the RQ is 1. The RQ for protein is 0.8.

449. The answer is D. *(Stryer, pp 76-77.)* The curve representing the binding of oxygen to hemoglobin as a function of the partial pressure of oxygen is sigmoidal in shape. A shift-to-the-right indicates a decrease in the capacity of hemoglobin to bind oxygen. Lowering the pH reduces the affinity of hemoglobin for oxygen, as does increasing the partial pressure of CO_2 at constant pH. The increased concentrations of H^+ and CO_2 therefore encourage the release of oxygen by hemoglobin. This is known as the Bohr effect.

450. The answer is D. *(White, ed 6. pp 86-87, 1022.)* According to the Henderson Hasselbalch equation, pH = pKa + log $\frac{base}{acid}$. In the case of 0.2 M lactate and 0.02 M lactic acid as presented in the question, pH = 3.9 + log 10 = 4.9.

451. The answer is E. *(White, ed 6. pp 1033-1034.)* In metabolic acidosis, blood bicarbonate is found to be low. In a situation of metabolic acidosis where bicarbonate has been consumed, compensation can be made by the respiratory mechanism of hyperventilation (sometimes referred to as "blowing off CO_2"), whereby pH is restored to normal by reducing plasma carbon dioxide.

452. The answer is B. *(White, ed 6. p 1093.)* The ATPase which hydrolyzes ATP, thereby releasing the energy that allows for muscle contraction, is located on the myosin molecule. The other components of muscle listed in the question lack ATPase activity.

453. The answer is C. *(McGilvery, pp 660-661.)* The daily intake of 62.5 g of high quality protein is above the minimum daily requirement for a 70 g adult (45 g protein per day). As the obligatory nitrogen losses are covered by the dietary intake, the individual will be in nitrogen balance (i.e., 0) and nitrogen loss will equal nitrogen intake.

454. The answer is A. *(White, ed 6. pp 296-300.)* Cyclic AMP phosphodiesterase is the intracellular enzyme that degrades cyclic AMP to $5'$-AMP. Adenylate cyclase synthesizes cyclic AMP from ATP. The remaining compounds listed in the question inhibit phosphodiesterase and would increase the levels of cyclic AMP.

455. The answer is E. *(Mahler, ed 2. p 38. McGilvery, pp 265-266.)* Muscular contractions are accompained by the conversion of ATP to ADP. However, the increase in ADP levels is dampened by the action of creatine kinase which maintains the following equilibrium:

$$\text{ADP + creatine phosphate} \rightleftharpoons \text{ATP + creatine}$$

In one gram of resting muscle there are about 40 micromoles of creatine + creatine phosphate, with about two-thirds in the phosphorylated form, and about seven micromoles of adenosine phosphates with nine-tenths present as ATP. Cyclic AMP concentrations in a similar muscle would be several orders of magnitude lower than the ATP concentration.

456. The answer is C. *(Mahler, ed 2. p 13.)* In any fluid, the maximum buffering action is achieved by the acid whose pKa most nearly approximates the pH of the fluid. Physiologic pH is about 7.4, so that among those buffers listed in the question, NaH_2PO_4 is the most effective.

457. The answer is B. *(Lehninger, ed 2. pp 42-43.)* Water molecules have a dipole nature and dissolve salts because of attractions between the water dipoles and the ions that exceed the force of attraction between the oppositely-charged ions of the salt. In addition, the latter force is weakened by the high dielectric constant of water. Non-ionic, but polar, compounds are dissolved in water because of hydrogen-bonding between water molecules and groups such as alcohols, aldehydes, and ketones.

458. The answer is C. *(Harper, ed 16. p 570)* The osmotic pressure due to blood proteins (the so-called oncotic pressure), which is normally about 25 mm Hg, is a function mainly of plasma albumin content. The normal distribution of water between the circulation and interstitial space is highly dependent upon a normal plasma albumin level. If this decreases, as it may in prolonged protein malnutrition or in liver disease, there will be an accumulation of water in the tissues (edema) with concomitant changes in the circulation.

459. The answer is D. *(Harper, ed 16. p 570.)* Albumin is responsible for 75 to 80 percent of the osmotic effect of plasma proteins. This is because it is the smallest and most abundant of the plasma proteins, though by weight it usually only comprises slightly more than half of the plasma proteins.

460. The answer is B. *(Mahler ed 2. pp 690-701.)* In tightly coupled mitochondria, respiration can proceed only in the presence of phosphate plus phosphate acceptor (ADP), as well as respiratory substrate. 2,4-Dinitrophenol uncouples respiration from phosphorylation, thus allowing active respiration in the absence of ADP. Oligomycin inhibits the ATPase responsible for catalyzing the phosphorylation of ADP. In mitochondria previously uncoupled by 2,4-dinitrophenol, oligomycin does not affect respiration.

461. The answer is A. *(White, ed 6. p 916.)* The secretion of the prostate gland contains very high concentrations of acid phosphatase. In **metastatic** prostatic carcinoma, the serum acid phosphatase levels usually are elevated, whereas in a tumor confined to the prostate, acid phosphatase levels may well be normal. While serum alkaline phosphatase levels also may be increased in carcinoma of the prostate, such increases arise in almost any situation involving tissue or cell damage.

462. The answer is B. *(Lehninger, ed 2. p 30.)* Lipid micelles are stable in water when the polar heads of the lipid face outward in contact with water and the hydrophobic, nonpolar tails turn inward to exclude water.

463. The answer is C. *(Bernhard, Dev Biol 35 [1973]:83-96.)* Gluconeogenesis from certain substrates and the hydroxylation of phenylalanine can be performed by the mammalian kidney as well as liver, and glycogen synthesis and storage are carried out by muscle as well as liver. The synthesis of serum albumin, however, is a uniquely hepatic function. A wide variety of cell types is sensitive to epinephrine.

464. The answer is B. *(Metzler, p 731.)* The uptake of exogenous cholesterol by cells results in a marked suppression of endogenous cholesterol synthesis. Low-density human lipoprotein not only contains the greatest ratio of bound cholesterol to protein, but also has the greatest potency in suppressing endogenous cholesterogenesis.

465. The answer is D. *(Gilbert, Proc Natl Acad Sci 72 [1975]:263-267. White, ed 6. p 618.)* In Sandhoff's disease, both the A and B forms of hexosaminidase are absent. Hexosaminidase A-deficiency is seen in Tay-Sachs disease. The other deficiency states described in the question are not currently identified with specific syndromes.

466. The answer is A. *(White, ed 6. p 779.)* The Lesch-Nyhan syndrome is a sex-linked, recessive disorder. Maple syrup urine disease is autosomal recessive, and familial hypercholesterolemia is codominant. Infantile autism and Cushing's syndrome do not have an inherited transmission.

467. The answer is A. *(Bondy, ed 7. p 805.)* Porphobilinogen is colorless and so has no effect on the color of urine when first excreted. On standing at an acid pH, urine containing colorless porphobilinogen may polymerize and oxidize to form porphyrins and other pigments, and so darken.

468. The answer is C. *(White, ed 6. pp 101-103.)* There are two conditions to be satisfied in the Donnan equilibrium: 1) electroneutrality of solutions on both sides of the membrane; and 2) chemical gradients for oppositely charged permeations must be equal and opposite. Thus, on the left side $(Na)_l = (Prot) + (Cl)_l = 100 + (Cl)_l$. On the right side, $(Na)_r = (Cl)_r$. Furthermore, $(Na)_l/(Na)_r = (Cl)_r/(Cl)_l$. Finally, according to conservation of mass, $(Na)_l + (Na)_r = 200$ and $(Cl)_l + (Cl)_r = 100$. All these conditions are satisfied for the following:

$$(Na)_l = 133\,⅓\,;\ (Na)_r = 66\,⅔\,;\ (Cl)_l = 33\,⅓\,;\ (Cl)_r = 66\,⅔$$

469. The answer is D. *(Bondy, ed 7. pp 123, 316-317.)* Muscle phosphorylase deficiency leads to a glycogen storage disease (McArdle's syndrome) and, in young adults, an inability to do strenuous physical work because of muscular cramps resulting from ischemia. The compromised phosphorylation of muscle glycogen characteristic of McArdle's syndrome compels the muscles to rely upon auxiliary energy sources like free fatty acids and ambient glucose.

470. The answer is C. *(White, ed 6. pp 1164-1168.)* Alkaline phosphatase, which is present in unusually high activity in osteoblasts, is thought to play a role in producing a local increase in the phosphate concentration. Phosphorylase kinase is involved in glycogenolysis and glucose 6-phosphate dehydrogenase is involved with the phosphogluconate pathway. Acid phosphatase is seen chiefly in the prostate, whereas ATPase is a widespread enzyme involved in membrane transport and a variety of other reactions.

471. The answer is A. *(Bondy, ed 7. p 596.)* In phenylketonuria, the enzyme phenylalanine hydroxylase, which synthesizes tyrosine from phenylalanine, is absent. This leads to a classic tetralogy of eczema, light pigmentation, seizures, and mental retardation, which can be prevented by early detection of the enzyme deficiency.

472. The answer is B. *(Bondy, ed 7. pp 314-315.)* Von Gierke's disease (type I glycogen storage disease) results from a deficiency of glucose 6-phosphatase. Phosphoglucomutase deficiency is seen in type VII glycogen storage disease. The other enzymes cited in the question are not specifically involved in the glycogenoses.

473. The answer is E. *(White, ed 6. pp 498, 876-877.)* Phenylketonuria results from phenylalanine hydroxylase deficiency. Tay-Sachs disease is caused by accumulation of excess gangliosides due to enzymatic deficiencies. Glucose 6-phosphatase deficiency leads to von Gierke's disease, and enzyme deficiencies result in accumulation of glucocerebrosides in Gaucher's disease. Caisson disease, or decompression sickness, is unrelated to an enzyme defect or deficit.

474. The answer is C. *(Bondy, ed 7. pp 314-316.)* Type IV glycogen storage disease (Andersen's disease) results from a lack of the branching enzyme amylo-$(1,4 \rightarrow 1,6)$-transglycosylase. The defect results in an abnormal glycogen structure with very long unbranched chains that can be detected on liver biopsy. Hepatomegaly and cirrhosis occur in this condition, which is usually fatal before the age of three years.

475. The answer is D. *(White, ed 6. pp 1044-1046.)* Parietal cells in the gastric mucosa secrete hydrochloric acid, lowering the gastric contents to a pH of 1 to 2. The acidity of gastric juice is neutralized in the intestine by alkaline pancreatic and biliary juice to a pH of 7 to 8. The pH of plasma is normally 7.4, while sweat is composed largely of half-normal saline, at a pH of 4.5 to 7.5.

476. The answer is C (2, 4). *(Mahler, ed 2. pp 489-491.)* Although the same intermediates may appear in both anabolic and catabolic pathways, one path is not simply the reverse of the other, because irreversible enzymatic steps often occur in the beginning of the reaction sequence. The same enzymes often appear in innumerable metabolic pathways.

477. The answer is A (1, 2, 3). *(Stryer, pp 85-89.)* 2,3-Diphosphoglycerate (2,3-DPG) and carbon dioxide both decrease the affinity of hemoglobin for oxygen by combining with, and stabilizing, the deoxygenated form of hemoglobin. Deoxyhemoglobin binds one molecule of 2,3-DPG on the symmetry axis in the central cavity. The reduction in affinity of hemoglobin for oxygen allows an adequate release of oxygen to the tissues. Carbon dioxide binds to the uncharged terminal amino groups of deoxyhemoglobin and is transported to the lungs. This binding of CO_2 also favors the deoxy-form of hemoglobin. An increase in hydrogen ion concentration changes the charge on histidine residues in hemoglobin, favoring the release of oxygen (the Bohr effect). An increase in hydroxyl ion concentration would mean a decrease in hydrogen ion concentration and an increase in the affinity of hemoglobin for oxygen.

478. The answer is D (4). *(Harper, ed 16. pp 334-336.)* During starvation, the level of gluconeogenic enzymes increases in the liver. Enzymes of the pentose cycle, glycolysis, and lipogenesis decrease in activity during starvation, as would be expected, since substrates for these pathways are then lacking.

479. The answer is C (2, 4). *(White, ed 6. pp 671, 1348, 1357-1358.)* Biotin and dehydroshikimic acid are not synthesized in mammalian tissue. While inositol is required in the diet, isotopic evidence suggests that at least small amounts are synthesized in mammalian tissue. Pathways also exist in mammalian tissues for the synthesis of choline via phosphatidyl choline.

480. The answer is E (all). *(Thorn, ed 8. pp 1580-1581.)* Alkaline phosphatase is a ubiquitous human intracellular enzyme. Any conditions that produce significant tissue damage will lead to increased serum alkaline phosphatase levels, as the enzyme is released by damaged cells. Thus, all four examples of tissue or cell damage listed in the question will be associated with increased levels of serum alkaline phosphatase.

481. The answer is A (1, 2, 3). *(White, ed 6. p 1001.)* While erythrocytes can metabolize glucose anaerobically to lactate via the glycolytic pathway, this process does not involve any net oxidation of glucose. The NADH derived from the triose phosphate dehydrogenase step is used to reduce pyruvate to lactate. Liver and heart are widely adaptable in their metabolic requirements, and can oxidize glucose well. The brain's cells utilize glucose and ketone bodies for oxidative metabolism and energy production.

482. The answer is C (2, 4). *(McGilvery, pp 272-280.)* The enzymatic capability to convert noncarbohydrate precursors to glucose is found in only three tissues — liver, kidney, and small intestine. The major site of gluconeogenesis is liver, although in starvation or in metabolic acidosis the contribution by the kidney increases.

483. The answer is B (1, 3). *(White, ed 6. pp 740-743.)* In the only reported case of tyrosinosis, the metabolic abnormality appeared to be a failure of the oxidase catalyzing the conversion of *p*-hydroxyphenylpyruvate to homogentisic acid, which is the normal pathway for the catabolism of phenylalanine and tyrosine. The block resulted in increased excretion of *p*-hydroxyphenylpyruvate, but not in an increase in its reduction to *p*-hydroxyphenyllactate.

484. The answer is D (4). *(Mahler, ed 2. p 857.)* Phenylketonuria is thought to result from a genetic deficiency in the enzyme phenylalanine hydroxylase, which converts phenylalanine to tyrosine. This disorder, which results in dementia, seizures, and dermatologic manifestations, is largely preventable by adequate perinatal screening and treatment.

485. The answer is B (1, 3). *(Bondy, ed 7. pp 578-579.)* Maple syrup urine disease results from a deficiency of the α-ketoacid dehydrogenases responsible for oxidation of the α-keto analogs of the branched chain amino acids (leucine, isoleucine, and valine). Consequently, these α-ketoacids accumulate in blood, urine, and spinal fluid, producing the biochemical abnormalities and odor characteristic of this disorder.

486. The answer is C (2, 4). *(Bondy, ed 7. pp 618, 620.)* On standing, urine containing porphobilinogen and homogentisic acid will turn black, as observed in acute intermittent porphyria and ochronosis, respectively. Metanephrine and 5-hydroxyindoleacetic acid, being colorless, will not discolor urine.

487. The answer is A (1, 2, 3). *(Stryer, pp 380-385.)* Mobilization of glycogen stores to produce glucose in the liver requires the phosphorolysis of the glycogen chain by the enzyme phosphorylase, the hydrolysis of α-1,6-glycosidic bonds by amylo-1,6-glucosidase (also known as the debranching enzyme), and the hydrolysis of glucose 6-phosphate derived from glucose 1-phosphate (product of phosphorylase) by glucose 6-phosphatase to produce glucose for export. A deficiency in any of these enzymes would compromise the ability of liver to mobilize glycogen stores. Amylo-(1,4 → 1,6)-transglycosylase functions in the separate pathway for glycogen synthesis.

488. The answer is E (all). *(Bondy, ed 7. pp 251, 262-263.)* All the afflictions mentioned in the question are known to be associated with abnormal glucose tolerance test curves. Mechanisms that promote glucose intolerance in these cases can include post-surgical rapid transit in the gastrointestinal tract, steroid therapy, and ectopic ACTH production.

489. The answer is B (1, 3). *(Lynch, ed 2. p 387.)* Once chronic insulin therapy is begun, serum insulin levels cannot be measured accurately because of the development of antibodies in response to immunoassay. Insulin requirements may decrease to nothing for a short period, a situation termed the "honeymoon phase." Impotence is a variable, but usually late, complication associated with diabetic neuropathy. Insulin requirement is governed by diet which, therefore, is the mainstay of "regulation" in the insulin-requiring, as well as in the noninsulin-requiring, diabetic patient.

490. The answer is E (all). *(White, ed 6. pp 605-606.)* The occurrence of poly-neuropathy or of the Wernicke-Korsakoff syndrome is probably due in large part to thiamine deficiency resulting from inadequate nutrition associated with chronic alcoholism. Metabolism of ethanol by alcohol dehydrogenase in the liver leads to high levels of NADH which in turn can lead to an increased conversion of pyruvate with a concomitant fall in blood glucose level, as well as a mild lactic acidosis. Increased mobilization and utilization of fat for metabolic fuel can augment the acidosis by increased ketone body formation and excretion. The excessive mobilization of fat can lead to fatty hepatosis (fatty liver) but the exact mechanisms involved are not entirely clear.

491. The answer is C (2, 4). *(McGilvery, pp 369-373, 526-527.)* Ketone bodies can be oxidized to carbon dioxide in virtually all tissues that contain mitochondria (such as brain, kidney, heart, skeletal muscle), but not in the liver. Cells without mitochondria, such as erythrocytes, cannot oxidize ketone bodies.

492. The answer is D (4). *(White, ed 6. pp 602-603.)* In the normal cycle of feed and fast, ketone body production is minimal; however, liver ketogenesis increases rapidly in starvation, or under conditions of unregulated release of free fatty acids from adipose tissue, as seen in insulin-dependent diabetes. Excess acetyl CoA from the oxidation of fatty acids in liver mitochondria is channelled through acetoacetyl CoA and 3-hydroxy 3-methyl glutaryl CoA to acetoacetate and β-hydroxybutyrate. These compounds are known as "ketone bodies," although the latter is not a ketone. Ketone bodies are not oxidized in the liver but released into the circulation, providing an important source of metabolic fuel to skeletal muscle, heart, and brain. Extrahepatic production of ketone bodies is minimal.

493-496. The answers are: 493-A, 494-C, 495-E, 496-D. *(Mahler, ed 2. p 805. White, ed 6. pp 877-878, 911, 990, 1001-1002.)* An erythrocyte glycolytic enzyme, pyruvate kinase is essential to pyruvate formation and its deficiency leads to severe hemolytic anemia.

The heme protein tryptophan pyrrolase, a ubiquitous dioxygenase, is a copper-containing enzyme whose deficiency is associated with Hartnup's disease. The disorder is autosomal recessive and, although sometimes asymptomatic, generally produces a dermatologic rash, cerebellar ataxia, and a wide range of psychiatric manifestations.

Lack of the hepatic enzyme uridine diphosphate glucuronate transferase is associated with the Crigler-Najjar syndrome. In this exceedingly rare disorder, the liver is unable to conjugate bilirubin and the resulting jaundice, particularly in affected newborns, may incur severe brain damage.

Wilson's disease, another rare inherited disorder, is characterized by a significant reduction in another copper-containing enzyme, ceruloplasmin. Its absence or deficit in the plasma is responsible for excessive hepatic and brain levels of copper. Wilson's disease, primarily an affection of childhood and youth, is thus characterized by liver damage and by neurologic (tremors) rather than psychiatric manifestations.

Tyrosinase deficiency, by exclusion, is regarded as responsible for albinism. In this disorder, an absence of melanin—the pigment responsible for the color of hair, skin, and retina—arises from a failure in tyrosine-to-melanin conversion, for which tyrosinase is required.

497-500. The answers are: 497-A, 498-A, 499-A, 500-D. *(Bondy, ed 7. pp 364-367.)* Endogenous insulin production is absent in patients with juvenile diabetes to a degree not seen in adult onset diabetes; consequently, such individuals have an absolute need for insulin. These patients have a greater tendency to develop ketoacidosis, and their diabetes is considerably more unstable and "brittle" than is the case with adult onset diabetes. Insulin administration to diabetic patients over the course of several months causes the development of antibodies to insulin. However, insulin resistance is rare in both juvenile and adult onset forms.

Bibliography

Beeson, P.B., and McDermott, W., eds. *Textbook of Medicine.* 14th ed. Philadelphia: Columbia Broadcasting System, W.B. Saunders Co., 1975.

Bernhard, H.P.; Darlington, G.J.; and Ruddle, F.H. "Expression of liver phenotypes in cultured mouse hepatoma cells: synthesis and secretion of serum albumin." *Developmental Biology* 35 (1973):83-96.

Bloom, W., and Fawcett, D.W. *A Textbook of Histology.* 10th ed. Philadelphia: Columbia Broadcasting System, W.B. Saunders Co., 1975.

Bondy, P.K., and Rosenberg, L.E., eds. *Duncan's Diseases of Metabolism.* 7th ed. Philadelphia: Columbia Broadcasting System, W.B. Saunders Co., 1974.

Davis, B.D., et al. *Microbiology.* 2nd ed. Hagerstown: Harper & Row Publishers Inc., 1973.

Gilbert, F.; Kucherlapati, R.; Creagan, R.P.; et al. "Tay-Sachs' and Sandhoff's diseases: the assignment of genes for hexosaminidase A and B to individual human chromosomes." *Proceedings of the National Academy of Sciences of the United States of America* 72 (1975):263-267.

Goodhart, R.S., and Shils, M.E., eds. *Modern Nutrition in Health and Disease: Dietotherapy.* 5th ed. Philadelphia: Lea & Febiger, 1974.

Harper, H.A., et al. *Review of Physiological Chemistry.* 16th ed. Los Altos: Lange Medical Publications, 1977.

Hayes, W. *The Genetics of Bacteria and Their Viruses.* 2nd ed. New York: John Wiley & Sons Inc., Halsted Press, 1976.

Hood, L.E.; Wilson, J.H.; and Wood, W.B. *Molecular Biology of Eucaryotic Cells.* Reading: Addison-Wesley Publishing Co., Benjamin-Cummings Publishing Co., 1975.

Jawetz, E.; Melnick, J.L.; and Adelberg, E.A. *Review of Medical Microbiology.* 13th ed. Los Altos: Lange Medical Publications, 1978.

Lehninger, A.L. *Biochemistry: The Molecular Basis of Cell Structure and Function.* 2nd ed. New York: Worth Publishers Inc., 1975.

Lewin, B. *Bacterial Genomes.* Gene Expression, vol 1. New York: John Wiley & Sons Inc., 1974.

Lindros, K.O.; Oshino, N.; Parrilla, R.; et al. "Characteristics of ethanol and acetaldehyde oxidation on flavin and pyridine nucleotide fluorescence changes in perfused rat liver." *Journal of Biological Chemistry* 249 (1974): 7956-7963.

Lynch, M.J., et al. *Medical Laboratory Technology and Clinical Pathology.* 2nd ed. Philadelphia: Columbia Broadcasting System, W.B. Saunders Co., 1969.

Mahler, H.R., and Cordes, E.H. *Biological Chemistry.* 2nd ed. Hagerstown: Harper & Row Publishers Inc., 1971.

McGilvery, R.W. *Biochemistry: A Functional Approach.* Philadelphia: Columbia Broadcasting System, W.B. Saunders Co., 1970.

Metzler, D.E. *Biochemistry: The Chemical Reactions of Living Cells.* New York: Academic Press Inc., 1977.

Miller, J.H. *Experiments in Molecular Genetics.* New York: Cold Spring Harbor Laboratory, 1972.

Sawin, C.T. *Hormones: Endocrine Physiology.* Boston: Little, Brown & Co., 1969.

Stryer, L. *Biochemistry.* San Francisco: W.H. Freeman & Co., 1975.

Thorn, G.W., et al., eds. *Harrison's Principles of Internal Medicine.* 8th ed. New York: McGraw-Hill Book Co., Blakiston Publications, 1977.

Watson, J.D. *Molecular Biology of the Gene.* 3rd ed. Reading: Addison-Wesley Publishing Co., Benjamin-Cummings Publishing Co., 1976.

White, A., et al. *Principles of Biochemistry.* 6th ed. New York: McGraw-Hill Book Co., 1978.

Williams, R.H., ed. *Textbook of Endocrinology.* 5th ed. Philadelphia: Columbia Broadcasting System, W.B. Saunders Co., 1974.

Wolfe, S. *Biology of the Cell.* Belmont: Wadsworth Publishing Co. Inc., 1972.